PROCEEDINGS
OF
THE JOHNS HOPKINS WORKSHOP
ON
CURRENT PROBLEMS IN PARTICLE THEORY
7

PREVIEWS JOHNS HOPKINS WORKSHOP PROCEEDINGS

1	BALTIMORE '74	....
2	BALTIMORE '78	....
3	FLORENCE '79	....
4	BONN '80	Lattice Gauge Theories. Integrable Systems.
5	BALTIMORE '81	Unified Field Theories and Beyond.
6	FLORENCE '82	Lattice Gauge Theories. Supersymmetry and Grand Unification.

The Johns Hopkins Workshops on Current Problems in Particle Theory are organized by the following universities:

UNIVERSITY OF BONN
UNIVERSITY OF FLORENCE
THE JOHNS HOPKINS UNIVERSITY

Organizing Committee:

Klaus DIETZ (Bonn)
Vladimir RITTENBERG (Bonn)

Roberto CASALBUONI (Florence)
Luca LUSANNA (Florence)
Gabor DOMOKOS (Johns Hopkins)
Susan KOVESI-DOMOKOS (Johns Hopkins)

PROCEEDINGS OF THE JOHNS HOPKINS WORKSHOP

ON

CURRENT PROBLEMS IN PARTICLE THEORY 7

BONN, 1983

(Bad Honnef June 21-23)

L A T T I C E   G A U G E   T H E O R I E S

*

S U P E R S Y M M E T R Y   A N D   G R A N D   U N I F I C A T I O N

Edited by

G. DOMOKOS and S. KOVESI-DOMOKOS

WORLD SCIENTIFIC

World Scientific Publishing Co Pte Ltd
P O Box 128
Farrer Road
Singapore 9128

ISSN 0275-617-X

ISBN 9971-950-63-4
9971-950-62-6 pbk

Printed in Singapore by Richard Clay (S.E. Asia) Pte. Ltd.

FOREWORD

This is the seventh in a series of Workshops on Current Problems in Particle Theory. As in the past, the basic purpose of this Workshop has been to provide a forum for theoretical physicists to discuss the most important problems of current research in theoretical particle physics. The discussions centered around a few invited talks, reproduced in this volume. In their talks, the speakers both summarized the current state of the art and presented new results. The main topics have been: nonperturbative aspects of gauge theories, which these days can be best studied in the lattice approximation, and supersymmetric theories: it appears that these topics have remained in the forefront of interest among theorists. While supersymmetric theories offer the hope of finding a satisfactory unification scheme in which, ultimately, all forces and particles observed in Nature will naturally find their place, the study of lattice gauge theories is providing an insight into the mechanism responsible for the confinement of elementary constituents into the observed hadrons. Thus, one is able to test the currently accepted theory of strong interactions - Quantum Chromodynamics - in a regime not accessible to techniques based on perturbation theory. While progress in both fields has been impressive, it is clear that still much remains to be done in these areas. We hope that the publication of this volume will contribute to the progress of research in these fields. It is unfortunate that this Proceedings has to be published without one of the important contributions to the Workshop. The Editors were very sorry to learn that Professor K. Symanzik was unable to submit a written version of his talk for reasons entirely beyond his control. We hope to welcome him at future Workshops and listen to his enlightening talks and remarks.

This Workshop took place in the beautiful setting of Bad Honnef, near Bonn, in the house maintained by the German Physical Society. We wish to thank the Deutsche Forschungsgemeinschaft for their generous financial support which made this Workshop possible and the staff of the Hölterhoff Stiftung for their contribution to running the Workshop.

We also wish to thank Miss Barbara Dreyfus for her able and dedicated help in producing the manuscript of this Proceedings and to Dr. K.K. Phua and the staff of World Scientific Publishing Co. for their understanding and encouragement in producing this volume.

The Organizing Committee

CONTENTS

RECENT RESULTS ON SIMULATIONS OF LATTICE QCD WITH QUENCHED FERMIONS

Roberto Petronzio

CERN
CH-1211 Geneva 23
Switzerland

I will talk about two subjects: the analysis of the hadron spectrum[1)] and the measurement of the string tension[2)]. All the results have been obtained by simulating lattice QCD on a 10^3*20 lattice, where the longest direction is defined as the time.

The generalities about the calculation of hadron masses are well known[3)]. One looks at the correlation function at a certain space-time distance of two local operators O carrying definite quantum numbers. The behaviour of such correlations for large values of the time distance provides information on the value of the lowest-lying state coupled to the operator O. In formulae:

$$C(t) \equiv \int d^3x \, \langle \mathcal{O}(x)\, \mathcal{O}^+(0) \rangle \underset{\text{large } t}{\sim} A \cdot e^{-M_{\mathcal{O}} t} \tag{1}$$

Examples of the operators O are:

$$\bar\psi \gamma_5 \psi, \; \bar\psi \gamma_\mu \psi, \; \dots \qquad \text{for mesons}$$
$$\epsilon^{ABC} [\psi^A_\alpha (C\gamma_5)_{\alpha\beta} \psi^B_\beta] \psi^C_\delta, \dots \qquad \text{for baryons} \tag{2}$$

where ψ are the quark fields.

The "quenching" approximation is made throughout the whole calculation: one neglects the diagrams where fermion loops are created by gluons. The integration over the fermion fields appearing in the average of Eq. (1) for a fixed gauge field configuration then reduces to the calculation of the quark propagator in the presence of a given external background field. The correlation function of Eq. (1) becomes a combination of two (mesons) or three (baryons) propagators suitably projected in colour and spin by the operators O. Various different field configurations are produced by a Monte Carlo algorithm, with the probability distribution:

$$dP(A) = \frac{d[A]\, e^{-S(A)}}{\int d[A]\, e^{-S(A)}} \tag{3}$$

where S(A) is the action of the gauge fields and the Wick rotation to a Euclidean space-time has been made.

The final result of Eq. (1) is obtained by averaging over the collected gauge field configurations the values of the correlation of the two operators:

$$\int dP(A)\, \mathcal{O}(x;A)\, \mathcal{O}^{+}(0;A) = \langle \mathcal{O}(x)\, \mathcal{O}^{+}(0) \rangle \quad . \tag{4}$$

Table 1 contains a summary of information on the type of action used on the lattice, on the parameters chosen*) and on the techniques used for generating the sample of gauge field configurations and for calculating the fermion propagator in a given external field.

The results show a definite improvement with respect to those obtained at the same value of $\beta = 6/g^2$ on a $5^3 \times 10$ lattice[3]; in particular:

i) the fluctuations among the values of the masses estimated from different configurations by fitting the time behaviour in Eq. (1) are drastically reduced;

ii) the results for mesons are stable against different choices of the operators O (for example, $\bar{\psi}\gamma^{i}\psi$ or $\bar{\psi}\sigma^{0i}\psi$ for the ρ and $\bar{\psi}\gamma^{5}\psi$ or $\psi^{+}\gamma^{5}\psi$ for the pion.

Concerning fluctuations, they are generally expected to decrease as the volume increases[4]: this, however, could not explain the drastic difference appearing in Fig. 1, where the value of the ρ mass obtained from different gauge field configurations (taken every hundred Monte Carlo sweeps) for the case of the $5^3 \times 10$ (black points) and of the $10^3 \times 20$ (open points) lattices is given. There is, in fact, an extra effect due to the use of periodic boundary conditions for fermions[5]. These conditions allow for diagrams like the one in Fig. 2a, where one quark follows a path lying in the original lattice (O) while the second reaches the final point on the replica (R). Seen on a two-dimensional torus, the second path wraps around it as in Fig. 2b. These diagrams are spurious: in practice, their contribution is small if the phase given by the line integral of the gauge field over the wrapping path fluctuates among the different trajectories.

*) We have always used $\beta \equiv 6/g^2 = 6$. For the comparison of these results with those at a different β ($\beta = 5.7$), see the talk by Wallace at this meeting.

Table 1

size of the lattice	$10^3 \cdot 20$
Gauge action	Wilson
Fermion action	Wilson
$\beta \equiv 6/g^2$	6
No. of gauge field configurations	27
Link updating procedure	Metropolis (10 hits)
No. of sweeps spent for thermalization	1300
Interval between gauge field configurations used for mass estimates	100 sweeps
Calculation of the quark propagator	Gauss-Seidel
Values of K used	0.150, 0.1535, 0.155

For a single trajectory, such a line integral is essentially a Polyakov loop: the magnitude of the average $\langle L \rangle$ of such a loop oriented, say, in the z direction over the orthogonal volume (x, y, t points), governs the importance of the spurious diagrams*). One obtains

$$\langle L \rangle_{5^3 \cdot 10} \sim 100 \cdot \langle L \rangle_{10^3 \cdot 20} \quad . \tag{5}$$

On the large lattice such diagrams are then negligible, as are the large fluctuations induced by their presence[5].

Concerning the stability against the use of different operators the values of the ρ mass in lattice units for different values of K for two different operators are reported in Table 2. The consistency between the two sets of results is a signal that the asymptotic regime of "large t" indicated in Eq. (1) has been reached. In this regime, the lowest-lying state is supposed to dominate the correlation function, and the only dependence upon the choice of the operator O will appear in the constant A multiplying the exponential in Eq. (1). As an extra check, one can construct the quantity:

$$m(t) = -\ln\left[\frac{C(t+1)}{C(t)}\right] \quad . \tag{6}$$

If only one state dominates C(t), m(t) $\to$ constant; if higher excited states contribute, m(t) will be a decreasing function of t. For mesons, the asymptotic regime is reached for $t > 7$; for baryons, this happens only at the lowest value of K (highest quark mass) while for higher values some contamination from higher excited states seems to persist up to $t \sim 9$**).

Table 2

K	0.150	0.1535	0.155
$\bar{\psi}\gamma^i\psi$	0.64 ± 0.02	0.51 ± 0.02	0.46 ± 0.02
$\bar{\psi}\sigma^{oi}\psi$	0.66	0.53	0.49

---

*) The expectation value of loops oriented in the time direction is used as an order parameter for the analysis of the finite temperature deconfining transition in QCD[6].

**) The contamination from excited states seems to be the main cause of discrepancy between the $5^3 \times 10$ and $10^3 \times 20$ results.

All the results on mesons and baryons are summarized in Fig. 3, where the values of the hadron masses in lattice units are given as a function of the quark mass in lattice units. The full curves are fits to the points which assume a linear dependence on the quark masses of the baryon and of the square of meson masses. In this figure there are two unknowns: the value of the lattice spacing in physical units and the appropriate value of the quark mass. Two inputs are then needed to fix such quantities: if one takes the value of the masses of ρ and pion, one gets the scales in GeV which are reported close to the lattice units scales corresponding to $a^{-1} = 2$ GeV. In addition, the value of the quark mass that gives a good fit to the π-ρ mass splitting is $m_u = m_d$*) $\sim$ 7 MeV, a very small value, well outside the domain where the calculation is performed. As an output, one obtains

$$m_{Proton} = [1.15 \pm 0.05]\ \text{GeV}$$
$$m_{\Delta} = [1.33 \pm 0.08]\ \text{GeV} \quad . \tag{7}$$

The proton mass turns out to be rather high.

An alternative procedure is that of using as inputs the values of masses of hadrons which are made of strange quarks. Using the ϕ mass and that of a pseudoscalar η [by SU(3) mass formulae it would be $\sim$690 MeV] one gets:

$$a^{-1} = 2.1\ \text{GeV}$$
$$m_s \simeq 120\ \text{MeV} \quad \text{and} \quad M_{\Omega^-} = 1.65\ \text{GeV} \quad .$$

By adding the pion as an input, one can also estimate the masses of hadrons containing one light and one strange quark. The summary of inputs and outputs for the two procedures described above is given in Table 3.

The general pattern is very encouraging for mesons and for "heavy" baryons (i.e., made of strange quarks): light baryons come out too heavy !

However, for these states, the contamination from higher excited states might be more severe: a preliminary result obtained by using a different operator based on the non-relativistic SU(6) wave function for the proton, gives

$$m^{Proton}_{NONREL.} \sim 1.0\ \text{GeV} \quad . \tag{8}$$

---

*) Up and down quarks are taken to be degenerate.

Table 3

Inputs	Outputs
m_π, m_ρ	a^{-1} = (2 ± 0.2) GeV $m_u = m_d \sim 7$ MeV m_p = (1.15 ± 0.05) GeV m_Δ = (1.33 ± 0.08) GeV
m_ϕ, m_{η_s}	a^{-1} = 2.1 GeV $K_s = 0.155 \rightarrow m_s$ = 120 MeV M_Ω = 1.65 GeV
m_ϕ, m_{η_s} and m_π	m_ρ = (0.82 ± 0.04) GeV m_K = (0.45 ± 0.03) GeV m_{K*} = (0.90 ± 0.04) GeV m_P = (1.2 ± 0.1) GeV m_Δ = (1.35 ± 0.15) GeV
$K_{critical}$ = 0.1570 ± 0.0003	

This value is certainly inconsistent with the previous one beyond the statistical fluctuations, and strongly suggests a systematic error due to the presence of radial excitations. A complete study of the proton correlation function using different operator is under way[7]. Another effect which might justify the high values obtained for baryons is the suppression, due to the quenching approximation, of diagrams like the one in Fig. 4, responsible for the corrections to the proton mass due to the emission and reabsorption of pions.

The second subject of the talk is a new measurement of the string tension K made on the same gauge field configurations from which the values of hadron masses have been extracted[2]. Indeed, the value of a^{-1} = 2 GeV obtained from the hadron mass fit would lead to a value for $\Lambda_{lattice}$, which, given the usually quoted relation between $\sqrt{K}$ and Λ_L ($\sqrt{K} \cong 170\Lambda_L$)[8], would imply $\sqrt{K} \sim 800$ MeV against the expected value of 400 ÷ 450 MeV. On the other hand, the existing estimates of K for SU(3) have been made on symmetric and relatively small plaquettes ($\sim 4 \times 4$ or 5×5), and may give results not directly interpretable in terms of the $q\bar{q}$ potential.

We decided to study the correlation of two Polyakov loops as a function of the distance. The expected behaviour is

$$< P_z(0,0,0)\, P_z(0,0,T)> \sim e^{-L_z V(T)} \underset{\text{large } T}{\longrightarrow} e^{-L_z T\cdot K} \tag{9}$$

where $P_z(x,y,t)$ is a loop in the direction z and of length L_z calculated at the point of co-ordinates x, y, t. In our case $L_z = 10$ and, given the rather large area already involved for $T \sim 2\div3$, a straightforward measurement of the quantity in Eq. (9) based on the available statistics of the order of 27 configurations (to be multiplied by two, since P_z and P_y loops are used) would be heavily affected by statistical fluctuations. We have used a new method which increases the statistics by about a factor one hundred[2)]. I will briefly describe the idea for a spin system: let us consider, in the Ising model, the value of the spontaneous magnetization $<\sigma_i>$

$$<\sigma_i> = \frac{\int d\sigma_1 d\sigma_2 \ldots d\sigma_i \ldots \cdot \sigma_i\, e^{-\beta H\{\sigma\}}}{\int d\sigma_1 d\sigma_2 \ldots d\sigma_i \ldots \ldots e^{-\beta H\{\sigma\}}} \tag{10}$$

where $H\{\sigma\}$ is the Hamiltonian of the system. If $H\{\sigma\} = \Sigma_{ik}J_{ik}\sigma_i\sigma_k$, the integration over the ith spin σ_i can be performed exactly and gives:

$$<\sigma_i> = < \mathrm{tgh}\, \beta J_{ik} \sigma_k > \; . \tag{11}$$

The second operator has the same average value (it comes after an exact integration), but much less fluctuations: for example, at $\beta = 0$, σ_i still fluctuates while the second operator gives identically zero (the correct result). The method can be extended to the case of average values of more spin variables if they are statistically independent: the terms in the Hamiltonian containing one of them do not contain any of the others. This method is particularly useful in the case of the calculation of the Polyakov loops: each of the links appearing in the loop is statistically independent from the others with a Wilson-type action. This is also true for the correlations if they are taken at distance equal to or greater than two. To increase the statistics, we have measured the correlation between a point and a plane:

$$\Pi(t) = \frac{1}{L_x L_y L_t} \sum_{x,y} \sum_{x',y',t'} < P_z(x,y,t')\, P_z(x',y',t+t') > \; . \tag{12}$$

The resulting function of t is given in Fig. 5, where the curve represents a least squares fit to the data points of the form $\pi(t) = A\cosh[(10-t)\eta]$. The periodicity is enforced by the use of periodic boundary conditions. The value of η is: $\eta = 0.39 \pm 0.04$ which, by using the expression in Eq. (9) with $L_z = 10$, gives

$$Ka^2\Big|_{\beta=6} = 0.039 \pm 0.004$$

If we use the value of a extracted from the hadron spectrum, we get $K \sim 400$ MeV, now in agreement with the expected value. This result confirms that the value of the lattice spacing in physical units at $\beta = 6$ is about 0.1 Fermi and renormalizes previous results which were converted into physical units assuming $a_{\beta=6} \sim 0.2$ Fermi. For example, this changes the conventionally quoted value of the 0^{++} glueball mass which now moves to $\sim$1200 MeV and, more dramatically, increases the latent heat of fusion of hadrons into gluons at the deconfinement temperature $T_c \sim 370$ MeV to about 10 GeV/Fm^3. This might have far-reaching consequences on the possibilities of studying such a phase transition by accumulating high energy densities in heavy ion collisions.

REFERENCES

1) H. Lipps, G. Martinelli, R. Petronzio and F. Rapuano, CERN preprint TH.3548 (1983), to be published in Phys. Lett. B.

2) G. Parisi, R. Petronzio and F. Rapuano, CERN preprint TH.3596 (1983), to be published in Phys. Lett. B.

3) H. Hamber and G. Parisi, Phys. Rev. D27 (1983) 208;
F. Fucito, G. Martinelli, C. Omero, G. Parisi and F. Rapuano, Nucl. Phys. B210 (FS6) (1982) 407;
D. Weingarten, Indiana University preprint IUHET-82 (1982).

4) See, for example, the talk by G. Parisi at the Les Houches Workshop, (March 1983) and H.W. Hamber, E. Marinari, G. Parisi and C. Rebbi, CEN-Saclay preprint SPhT/83/54 (1983).

5) G. Martinelli, G. Parisi, R. Petronzio and F. Rapuano, Phys. Lett. 122B (1983) 283.

6) See, for example, T. Celik, J. Engels and H. Satz, Bielefeld preprint, BI-TP 83/07 (1983).

7) G. Martinelli, G. Parisi, R. Petronzio and F. Rapuano, in preparation.

8) R. Ardill, M. Creutz and K. Moriarty, Phys. Rev. D27 (1983) 1956.

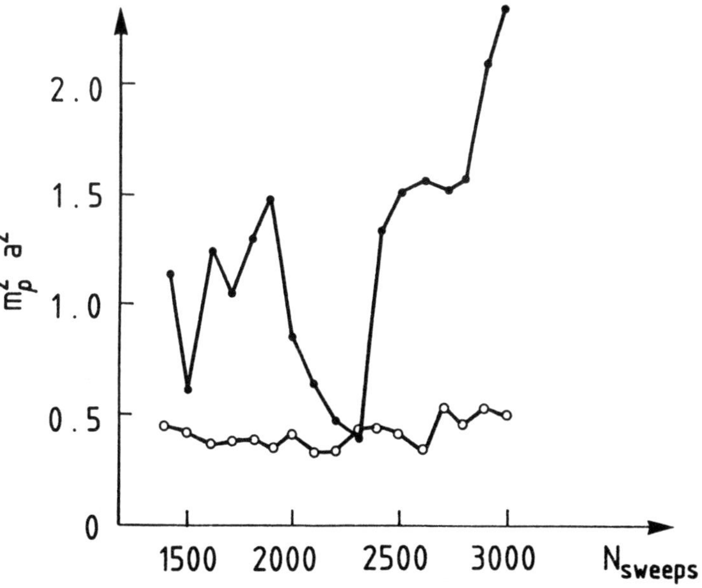

Fig. 1

Fig. 1: The fluctuations of the ρ mass squared in lattice spacing units at $K = 0.1475$ on the $5^3 \cdot 10$ lattice (closed points) and at $K = 0.150$ on the $10^3 \cdot 20$ lattice (open points).

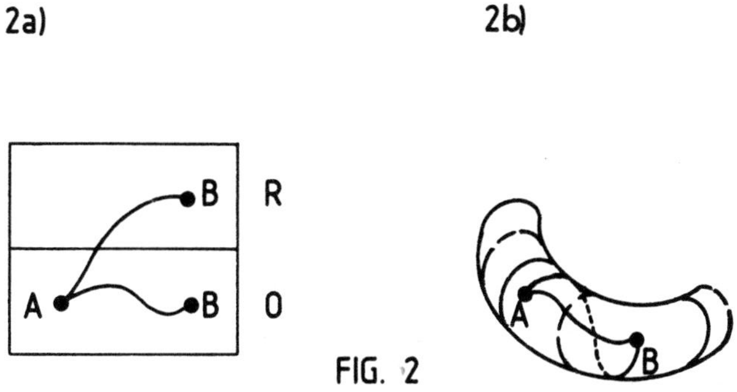

FIG. 2

Fig. 2: a) A spurious diagram where one quark trajectory lies in the original lattice (O) and the other goes into the replica (R).

b) The diagram of Fig. 2a seen on the torus.

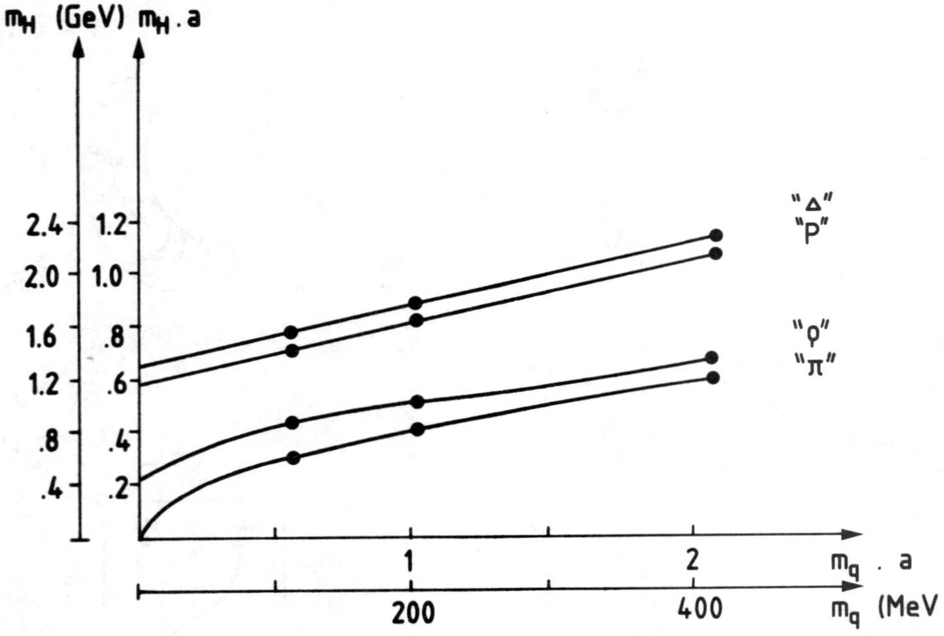

FIG. 3

Fig. 3: Various hadron masses in lattice spacing and in physical units against quark masses in lattice spacing and in physical units. The lines are the fits to the data points.

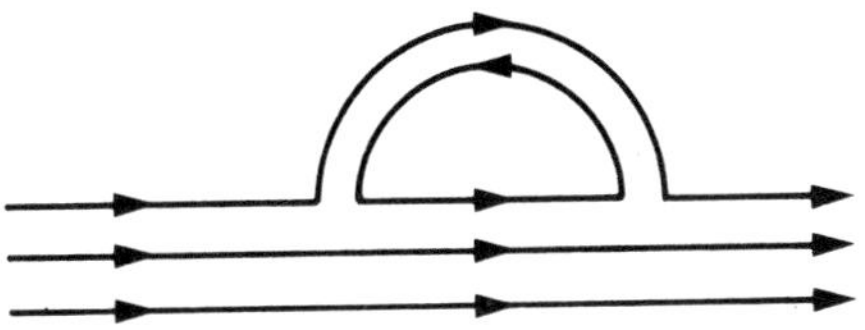

FIG. 4

Fig. 4: Example of a diagram for the proton propagation which is absent in the quenching approximation.

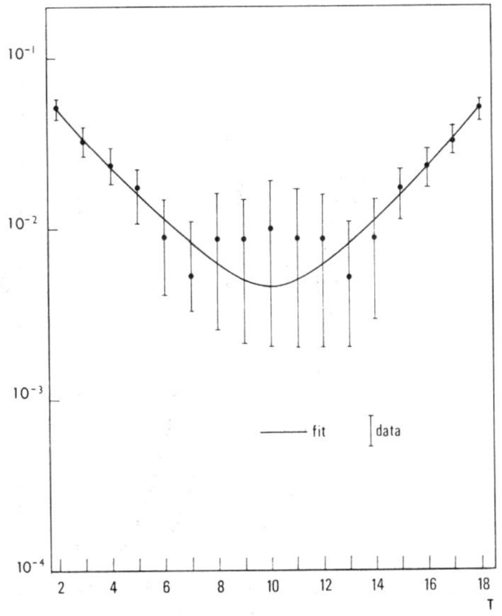

FIG. 5

Fig. 5: The correlation of Polyakov loops $\pi(t)$ (see text) as a function of the distance t.

THE DIRAC-KÄHLER LATTICE REGULARIZATION OF THE FERMIONIC DEGREES OF FREEDOM*)

by

Peter Becher

Physikalisches Institut der Universität Würzburg,
Am Hubland, D-8700 Würzburg, Germany (FRG)

1) Introduction

The Kähler equation is given by [1]

$$(d - \delta + m)\phi = 0. \tag{1.1}$$

It is defined on any pseudo-Riemannian manifold as well as on any smooth triangulation of such a manifold. On a manifold, ϕ is a general differential form, d is the exterior derivative, δ the generalized divergence and m a mass term. On the triangulation of a manifold, i.e. on a lattice, ϕ is a general co-chain, that is a linear functional on the linear combinations of the lattice elements of different dimension: points, links, plaquettes, ..., d is the dual boundary operator and δ the dual co-boundary operator.

Erich Kähler showed in 1961/62 [1,2] that for the flat space time manifold the Kähler equation (1.1) can be completely reinterpreted in terms of the free Dirac equation

$$(\gamma^\mu \partial_\mu + m)\psi = 0. \tag{1.2}$$

A first remark on the relation between these two equations is given by the fact that

$$d^2 = 0, \quad \delta^2 = 0. \tag{1.3}$$

It follows that the square of the Kähler operator is the Laplacian:

$$(d - \delta)^2 = -(d\delta + \delta d), \tag{1.4}$$

a property which $d - \delta$ shares with the Dirac operator $\gamma^\mu \partial_\mu$. The reinterpretation of the Kähler equation in terms of the Dirac equation remains true if an abelian or non-abelian gauge field is coupled minimally to ϕ and ψ, respectively.

In curved space time equ.'s (1.1) and (1.2) are inequivalent. In 1978 W. Graf proposed to use the Kähler equation as an alternative to the usual spinor concept [3]. In fact, in curved space time, one is enforced to assume tensor rather than spinor transformation properties of the Kähler field ϕ with respect to the local Lorentz gauge symmetry [4]. T. Banks, Y. Dothan and D. Horn discussed possible phenomenological consequences of this inequivalence [5]. B. Holdom modified the massless Kähler equation such that it describes a standard gauge theory of fermions plus gravity [6]. It has also been suggested [5,7] that the Kähler description of fermions might provide a starting point for their incorporation into Kaluza-Klein theories and a supersymmetric model in terms of Kähler fields was formulated [8].

The purpose of this talk is to review some of the applications of the Kähler equation for the description of fermions on a lattice. It turns out that the free Kähler equation is the formal continuum limit of free Susskind lattice fermions [9]. The situation for fields with gauge interaction is described below.

The concept how to translate the continuum Kähler equation to the lattice is purely geometric. It rests on an extended correspondence between the calculus of differential forms and lattice notions known from algebraic topology. The result is a tight analogy between the continuum and the lattice formulation of Kähler-Dirac fields. This makes it straightforward on the lattice to

- discuss symmetries and derive conserved currents,
- introduce gauge interactions
- quantize Kähler-Dirac fields and do f.i. weak coupling perturbation calculations.

The Kähler formulation might even open the possibility to invent new types of models which are inspired by geometric intuition. An example, which points in this direction, is the formulation of a minimal 4 flavour QCD model with non-degenerate quark masses which was given in ref. [10]. The mass term of this model was also discussed in ref. [11].

In the first part of this talk the lattice Kähler equation for free fermions is derived and discussed. The second part gives the connection with the Susskind reduction of the naive lattice Dirac equation and with other formulations of Kähler fermions on the lattice. In the third part I mention some of the attempts to incorporate gauge interactions into the geometric fermion concept and review the results on the chiral symmetry and chiral current anomaly of the gauged Kähler equation. The discussion is restricted from the very beginning to 4 dimensional flat euclidean space time and to a hypercubical lattice. Because the Kähler equation is formulated completely coordinate free, it is straightforward to extend the analysis to other space time dimensions and other lattice shapes. The latter may be of interest both for theoretical [12,13] and practical [14] reasons. For Kähler fermions one expects always the same number of fermionic degrees of freedom in the naive continuum limit. This has been demonstrated by M. Göckeler in a 2-dimensional model [15]. The different numbers of components of the lattice Kähler field have been listed by H. Raszillier for the different lattice holohedries in 2, 3 and 4 dimensions [16].

2) The Kähler equation for free fermions on the lattice

In a purely geometric setting this equation was derived and discussed in ref. [17]. I follow the lines of this paper.

a) The definition of the Kähler equation in cartesian coordinates in the continuum

With help of the cartesian coordinates $(x^\mu) \equiv (x^1, \dots, x^4)$ in the flat euclidean space time continuum, the standard basis of differential forms is given by the exterior or Grassmann products of 1-forms dx^μ :

$$dx^{\mu_1} \wedge \dots \wedge dx^{\mu_h} \equiv dx^H, \tag{2.1}$$

where H is the ordered index set $H = (\mu_1, \dots, \mu_h), \mu_1 < \dots < \mu_h$,

$dx^{\varnothing} \equiv 1$, $\varnothing$ the empty set. The Kähler field can be expanded with respect to this basis according to

$$\phi = \sum_{h=0}^{4} \frac{1}{h!} \varphi_{\mu_1 \dots \mu_h}(x)\, dx^{\mu_1} \wedge \dots \wedge dx^{\mu_h} \equiv \sum_{H} \varphi(x, H)\, dx^{H}. \tag{2.2}$$

This shows that the Kähler field can be viewed as a coherent superposition of antisymmetric tensor field:

$$\{ \mathring{\varphi}(x) \equiv \varphi(x, \varnothing),\ \varphi_{\mu}(x) \equiv \varphi(x, \{\mu\}),\ \varphi_{\mu\nu}(x) \equiv \varphi(x, \{\mu\nu\}), \dots \}.$$

The exterior differential $d\phi$ of ϕ is given by

$$d\phi = dx^{\mu} \wedge \partial_{\mu} \phi \tag{2.3}$$

in cartesian coordinates, where

$$\partial_{\mu} \phi := \sum_{H} (\partial_{\mu} \varphi(x, H))\, dx^{H} \tag{2.4}$$

means differentiation of the coefficient functions of the Kähler field. In order to define the generalized divergence δ, we introduce the $\star$-operator, which transforms every p-form into its orthogonal (4-p)-form; with the help of the totally antisymmetric tensor $\varepsilon_{\mu\nu\rho\sigma}$ this reads

$$\begin{aligned} \star 1 &= \frac{1}{4!} \varepsilon_{\mu\nu\rho\sigma}\, dx^{\mu} \wedge dx^{\nu} \wedge dx^{\rho} \wedge dx^{\sigma}, \\ \star dx^{\mu} &= \frac{1}{3!} \varepsilon^{\mu}{}_{\nu\rho\sigma}\, dx^{\nu} \wedge dx^{\rho} \wedge dx^{\sigma}, \\ \star dx^{\mu} \wedge dx^{\nu} &= -\frac{1}{2!} \varepsilon^{\mu\nu}{}_{\rho\sigma}\, dx^{\rho} \wedge dx^{\sigma} \quad \text{etc.}, \quad \star\star = \mathbb{1}. \end{aligned} \tag{2.5}$$

The divergence operator is then given by

$$\delta = -\star^{-1} d \star. \tag{2.6}$$

This finishes the definition of the Kähler equation (1.1) in cartesian coordinates.

b) Translation of the Kähler equation to the lattice

In order to write the Kähler equation (1.1) on a hypercubical lattice, I would like to remind you of some of the correspondences between continuum and lattice concepts:

The elements of the lattice are points x , links between x and $x+e_\mu$: $(x, \{\mu\})$, where e_μ is a free vector of the length of the lattice constant a in direction μ , plaquettes at x , extending in $\mu\nu$ -direction and so on; we introduce the multiindex notation (x, H) for these lattice elements, H an ordered subset of $\{1,\dots,4\}$, $x \equiv (x, \emptyset)$. The linear combinations of these lattice elements are called chains:

$$C = \sum_{x,H} c(x,H)\cdot(x,H). \tag{2.7}$$

If all coefficients $c(x,H)$ in equ. (2.7) are zero except those with cardinality of H equal to p, the chains are called p-chains. p-forms can be integrated over p-chains:

$${}^p\omega\,({}^pC) = \int_{{}^pC} {}^p\omega . \tag{2.8}$$

This defines a linear mapping from the space of p-chains into the underlying field, i.e. a p-co-chain. Co-chains are the natural lattice analogues of differential forms! We introduce the dual basis $d^{x,H}$ of the basis of chains (x,H):

$$d^{x,H}((x',H')) = \delta^{x}_{x'}\ \delta^{H}_{H'} . \tag{2.9}$$

A general co-chain can be expanded according to

$$\Phi = \sum_{x,H} \varphi(x,H)\, d^{x,H} . \tag{2.10}$$

This is the Kähler field in cartesian coordinates on the lattice. The different operations on forms can be translated in a similar way. We start with the exterior derivative. From equ. (2.8) and Stokes' theorem

$$(d\,{}^p\omega)({}^{p+1}C) = \int_{{}^{p+1}C} d\,{}^p\omega = \int_{\Delta\,{}^{p+1}C} {}^p\omega = {}^p\omega(\Delta\,{}^{p+1}C) \tag{2.11}$$

it follows that the corresponding operation on co-chains is the adjoint of the boundary operator Δ , the dual boundary operator $\breve{\Delta}$:

$$(\breve{\Delta}\, d^{x,H})((x',H')) = d^{x,H}(\Delta(x',H')). \tag{2.12}$$

Similarly, the $\star$ operation is the adjoint of a mapping between the lattice and the dual lattice. A point is sent into a hypercube centered around this point, a link is mapped into a 3-cube centered around the middle of the link and so on; translating the definition (2.6) of the divergence to the lattice in this way, one concludes that the lattice analogue of δ is the adjoint of the co-boundary operator ∇, the dual co-boundary operator $\breve{\nabla}$:

$$(\breve{\nabla}\, d^{x,H})((x',H')) = d^{x,H}(\nabla(x',H')). \tag{2.13}$$

The co-boundary of a point is thereby the sum of all links ending in that point, the co-boundary of a link is the sum of oriented plaquettes, for which this link lies in their boundary and so on. With the help of equ.'s (2.10, 12, 13) the Kähler equation on the lattice is given by

$$(\breve{\Delta} - \breve{\nabla} + m)\,\phi = 0, \tag{2.14}$$

or, explicitly, in cartesian coordinates:

$$\sum_{\mu\in H} \rho_{\{\mu\},H\setminus\{\mu\}} \nabla_\mu^+ \varphi(x, H\setminus\{\mu\}) + \sum_{\mu\notin H} \rho_{\{\mu\},H} \nabla_\mu^- \varphi(x, H\cup\{\mu\}) + m\,\varphi(x,H) = 0. \tag{2.15}$$

Here, $\rho_{H,K}$ is a sign function explained in ref. [17] and

$$
\begin{aligned}
(\nabla_\mu^+ \varphi)(x,H) &:= \frac{1}{a}\left(\varphi(x+e_\mu,H) - \varphi(x,H)\right),\\
(\nabla_\mu^- \varphi)(x,H) &:= \frac{1}{a}\left(\varphi(x,H) - \varphi(x-e_\mu,H)\right).
\end{aligned}
\tag{2.16}
$$

Equ. (2.15) is a system of linear homogeneous difference equations for the 16 coefficient functions φ . In a complete missinterpretation of E. Kähler's intention it is possible to discuss this system in a purely algebraic context. For this, introduce 16 × 16 matrices

$$
\begin{aligned}
(A_+^\mu)_{H,H'} &= \begin{cases} \rho_{\{\mu\},H\setminus\{\mu\}}\ \delta_{H'}^{H\setminus\{\mu\}} & \text{if } \mu\in H,\\ 0 & \text{if } \mu\notin H,\end{cases}\\
(A_-^\mu)_{H,H'} &= \begin{cases} 0 & \text{if } \mu\in H,\\ \rho_{\{\mu\},H}\ \delta_{H'}^{H\cup\{\mu\}} & \text{if } \mu\notin H.\end{cases}
\end{aligned}
\tag{2.17}
$$

Then the Kähler equation reads

$$
(A_+^\mu \nabla_\mu^+ + A_-^\mu \nabla_\mu^- + m)\,\varphi = 0 \tag{2.18}
$$

if we skip indices. The hypercomplex numbers $A_\pm^\mu$ generate the full matrix algebra of the 16 × 16 matrices; they satisfy the algebraic relations

$$
\{A_+^\mu, A_+^\nu\} = 0,\quad \{A_-^\mu, A_-^\nu\} = 0,\quad \{A_+^\mu, A_-^\nu\} = \delta^{\mu\nu}. \tag{2.19}
$$

In the form (2.18, 19) the Kähler equation was discussed in ref. [18] as the most general first order linear difference equation without spectrum degeneracy and nearest neighbour derivative approximation. The relations (2.19) characterize the hypercomplex numbers $A_\pm^\mu$ up to isomorphisms uniquely [18]. Equ. (2.17) is only one special representation (see equ.'s (3.26,28) below).

c) The energy-momentum spectrum of the lattice Kähler equation

In analogy to the continuum relations (1.3,4), the dual boundary and dual co-boundary operator are nilpotent:

$$\breve{\Delta}^2 = 0, \quad \breve{\nabla}^2 = 0 \tag{2.20}$$

such that the square of the Kähler operator on the lattice is the correct Laplacian, too:

$$(\breve{\Delta} - \breve{\nabla})^2 = -(\breve{\Delta}\breve{\nabla} + \breve{\nabla}\breve{\Delta}). \tag{2.21}$$

If we consider plane wave solutions on the lattice

$$\varphi(x, H) = u(p, H) \cdot e^{-ip_\mu x^\mu}, \quad -\frac{\pi}{a} < p_\mu \leq \frac{\pi}{a}, \tag{2.22}$$

the iterated Kähler equation becomes

$$\left(\sum_\mu \left(\frac{2}{a} \sin \frac{p_\mu a}{2}\right)^2 + m^2\right) u(p, H) = 0. \tag{2.23}$$

This shows, that there is no spectrum degeneracy in the 1. Brillouin zone $-\frac{\pi}{a} < p_\mu \leq \frac{\pi}{a}$, the number of degrees of freedom on the lattice is the same as in the continuum. In fact, J. M. Rabin [19] has given a topological argument why the geometrical correct translation of a continuum theory to the lattice gives no spectrum degeneracy.

d) The relation between the Kähler equation and the Dirac equation in the continuum

The relation between equ.'s (1.1) and (1.2) is given through an extension of the Grassmann algebra of differential forms to a Clifford algebra. For 1-forms the Clifford product "$\vee$" is defined according to

$$dx^\mu \vee \;=\; dx^\mu \wedge \;+\; e^\mu \lrcorner \,, \tag{2.24}$$

where $e^\mu \lrcorner$ is the derivation of forms with respect to dx^μ:

$$e^\mu \lrcorner\, 1 = 0\,, \qquad e^\mu \lrcorner\, dx^\nu = g^{\mu\nu}. \tag{2.25}$$

It turns out that the generalized divergence (2.6) can be written with the help of $e^\mu \lrcorner$:

$$\delta = -e^\mu \lrcorner\, \partial_\mu . \tag{2.26}$$

Therefore, the Kähler operator acts via Clifford multiplication:

$$d - \delta = dx^\mu \wedge \partial_\mu + e^\mu \lrcorner\, \partial_\mu = dx^\mu \vee \partial_\mu . \tag{2.27}$$

Since $dx^\mu \vee dx^\nu + dx^\nu \vee dx^\mu = 2g^{\mu\nu}$, the mapping

$$\gamma^\mu \longmapsto dx^\mu \vee$$

defines a representation of the algebra of γ-matrices on the 16-dimensional space of differential forms. Because of the Pauli theorem this representation can be decomposed into four 4-dimensional irreducible representations equivalent to those generated by the standard γ-matrices. A basis for the space of differential forms, where the decomposition is explicit is given by

$$Z = \sum_H (-1)^{\binom{h}{2}} (\gamma_H)^T dx^H, \quad \gamma_H = \gamma_{\mu_1} \cdots \gamma_{\mu_h} \quad \text{if } H = (\mu_1, \ldots, \mu_h). \tag{2.28}$$

The basis vectors Z_{ab} are specified by the two Dirac indices of the γ-matrices on the right hand side. The Kähler field can be expanded with respect to this Dirac basis:

$$\phi = \sum_{H} \varphi(x,H)\, dx^{H} = \sum_{a,b} \psi_a^{(b)}(x)\, Z_{ab} \,. \tag{2.29}$$

The relation between the cartesian coefficients $\varphi(x,H)$ and the Dirac coefficients $\psi_a^{(b)}(x)$ is

$$\varphi(x,H) = \mathrm{trace}\ \psi(x)(\gamma_H)^{+}\,, \quad \psi_a^{(b)}(x) = \frac{1}{4}\sum_{H} \varphi(x,H)(\gamma_H)_{ab} \,. \tag{2.30}$$

It follows that ϕ is a solution of the Kähler equation

$$(d - \delta + m)\phi \equiv (dx^{\mu} \vee \partial_{\mu} + m)\phi = 0 \tag{2.31}$$

if and only if for each $b = 1,\ldots,4$, $\psi^{(b)}$ is a solution of the Dirac equation

$$(\gamma^{\mu}\partial_{\mu} + m)\psi^{(b)} = 0 \,, \quad \psi^{(b)}(x) = \begin{pmatrix} \psi_1^{(b)}(x) \\ \vdots \\ \psi_4^{(b)}(x) \end{pmatrix} . \tag{2.32}$$

The Kähler equation in the continuum decouples into four Dirac equations. Following a suggestion of L. Susskind |9| for the equivalent lattice case, the index b is called the flavour index.

Is it possible to decouple the Kähler equation into Dirac equations on the lattice, too? For an answer it is convenient to interprete the reduction of the Kähler equation to Dirac equations in the continuum as the diagonalization of a symmetry group $\mathcal{R}$, the reduction group of the Kähler equation. This group is a subgroup of the (global) "flavour symmetry group", which acts on the flavour index in the Dirac basis $\psi^{(b)}$.

e) The flavour symmetry group of the Kähler equation

For a description of the flavour symmetry group one should distinguish between real and complex coefficients $\varphi(x,H)$ of the Kähler field.

If the $\varphi(x,H)$ are real, the charge conjugate Kähler field is ${}^c\phi = \phi \vee dx^2$. In Dirac components and in the Weyl basis of the γ-matrices this reads

$$ {}^c\psi^{(b)}(x) = \psi^{(b')}(x)\,(\gamma_2)_{b'}{}^{b}\;, \quad \gamma_2 = \begin{pmatrix} 0 & & & 1 \\ & & -1 & \\ & -1 & & \\ 1 & & & 0 \end{pmatrix}. \tag{2.33} $$

Therefore the Kähler equation describes two complex Dirac fields. In this case, the flavour group is the real orthogonal group $O(4)$. Although the real Kähler equation is the most natural one from the geometrical point of view there is up to now no discussion of a lattice model with interaction in the literature, which is based on the real Kähler equation.

Therefore, we consider here only the case of complex coefficients $\varphi(x,H)$. The flavour symmetry is given in differential forms by right-$\vee$-multiplication with constant differentials

$$ c = \sum_H c(H)\, dx^H . \tag{2.34} $$

The transformations

$$ \phi \mapsto \phi \vee c \tag{2.35} $$

of the Kähler field are symmetries of the Kähler equation. They constitute the global flavour group $\mathcal{U}(4)$ and transform the flavour components in the Dirac basis, because equ. (2.35) is equivalent to

$$ \psi_a^{(b)}(x) \mapsto \psi_a'^{(b)}(x) = \hat{C}^{b}{}_{b'}\, \psi_a^{(b')}(x) \tag{2.36} $$

with

$$\hat{C} = \sum_{H} c(H)\,(\gamma^{\mu}_{H})^{T}. \tag{2.36'}$$

The subsidiary conditions

$$\phi \vee \varepsilon = \pm \phi \;, \quad \phi \vee \tau = \pm \tau \tag{2.37}$$

characterize the four different flavour sectors $\psi^{(b)}$, $b = 1, \ldots, 4$, where $\varepsilon = dx^1 \wedge \ldots \wedge dx^4$ is the volume element and $\tau = i\,dx^1 \wedge dx^2$. This can be seen explicitly with the help of equ. (2.36'):

$$\hat{\varepsilon} \equiv \gamma^{\mu}_{5} = \begin{pmatrix} -1 & & & \\ & -1 & & 0 \\ & & +1 & \\ 0 & & & +1 \end{pmatrix} , \quad \hat{\tau} \equiv i\gamma^{\mu}_{1}\gamma^{\mu}_{2} = \begin{pmatrix} -1 & & & \\ & +1 & & 0 \\ & & -1 & \\ 0 & & & +1 \end{pmatrix} . \tag{2.38}$$

The subgroup $\mathcal{R} = \{1, \tau, \varepsilon, \varepsilon \vee \tau\} \simeq Z_2 \times Z_2$ of the flavour symmetry group is the desired reduction group of the Kähler equation in the continuum.

f) The complete reduction of the free lattice Kähler equation into Dirac equations

The reduction group $\bar{\mathcal{R}}$ on the lattice is generated by the constant cochains

$$\varepsilon = \sum_{x} d^{x,\{1,2,3,4\}} \;, \quad \tau = i \sum_{x} d^{x,\{1,2\}} \;, \tag{2.39}$$

which are the lattice versions of the corresponding forms in the continuum. It is true on the lattice that

$$\phi \mapsto \hat{\varepsilon}\phi := \phi \vee \varepsilon \;, \quad \phi \mapsto \hat{\tau}\phi := \phi \vee \tau \tag{2.40}$$

are symmetries of the Kähler equation. Here, "$\vee$" is a lattice analogue of the Clifford product (2.24) [17]. The group $\bar{\mathcal{R}}$, however, is of infinite order. It is intertwined with translations because

$$\hat{\varepsilon}^2 = T_{-\sum_\mu e_\mu} \quad \text{and} \quad \hat{\tau}^2 = T_{-(e_1+e_2)} \tag{2.41}$$

are translations by $(e_1 + \ldots + e_4)$ and $(e_1 + e_2)$, respectively. The reason for this is that the exterior product of co-chains is different from zero only if a matching condition is fulfilled (see fig. 1):

$$d^{x,H} \wedge d^{y,K} = \begin{cases} \rho_{H,K} \; \delta^{x+e_H, y} \; d^{x, H\cup K} & \text{if } H\cap K = \emptyset, \\ 0 & \text{otherwise.} \end{cases} \tag{2.42}$$

This carries over to the Clifford product.

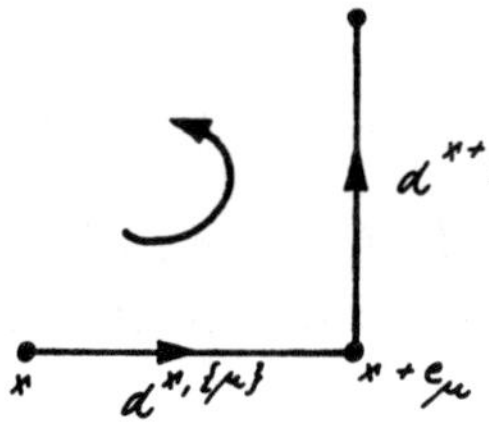

Fig. 1: The matching condition for $d^{x,\{\mu\}} \wedge d^{x+e_\mu,\{\nu\}} = d^{x,\{\mu\nu\}}$

Nevertheless, the symmetry group $\overline{\mathcal{R}}$ allows the reduction of the Kähler equation into Dirac equations on the lattice. The diagonalization of its translational subgroup,however, requires the transition to momentum space:

If $\phi = \sum_{x,H} \varphi(x,H)\, d^{x,H}$ satisfies the Kähler equation and if we write

$$\varphi(x,H) = e^{-i\frac{p}{2}e_H} \, \mathrm{trace}\,(\gamma_H^{\,+}\, \psi(p))\, e^{-ipx}, \quad e_H = \sum_{\mu\in H} e_\mu, \tag{2.43}$$

then the components $\psi^{(b)}(p)$ are solutions of the Dirac equation

$$\Big(\sum_\mu (-2i \sin \tfrac{p_\mu a}{2})\, \gamma_\mu + m\Big)\psi^{(b)}(p) = 0, \quad b = 1,\ldots,4. \tag{2.44}$$

This is the complete reduction of the lattice Kähler equation into Dirac equations.

The reduction of the Kähler equation in the continuum is possible also in the presence of a gauge interaction. The reduction group commutes with the gauge group. This is not the case on the lattice. Therefore the real problem of fermions on the lattice is a dynamical problem!

Before I come to gauge interactions, I would like to compare the Kähler description of fermions on the lattice with other approaches, which are equivalent to the Kähler formulation for free fields. I am not talking about Wilson's fermion description [20] and the SLAC fermions [21].

3) The Susskind reduction of the naive lattice Dirac equation

a) The naive Dirac equation on the lattice

The starting point of various fermion descriptions on the lattice is the so-called naive lattice Dirac equation:

$$(\gamma^\mu \nabla_\mu + m)\psi = 0, \tag{3.1}$$

where ∇_μ is the antisymmetric difference approximation

$$\nabla_\mu \psi(y) = \frac{1}{2b}(\psi(y+\varepsilon_\mu) - \psi(y-\varepsilon_\mu)) \tag{3.2}$$

of the derivative ∂_μ. The equation is defined on a hypercubic lattice with lattice points y and lattice constant b. For plane waves

$$\psi_\alpha(y) = u_\alpha(p)\, e^{-ipy}, \quad -\frac{\pi}{b} < p_\mu \leq \frac{\pi}{b}, \tag{3.3}$$

the iterated naive Dirac equation becomes

$$\Delta^{-1}(p)\, u_\alpha(p) \equiv \Big(\sum_\mu \big(\tfrac{1}{b}\sin p_\mu b\big)^2 + m^2\Big)\, u_\alpha(p) = 0. \tag{3.4}$$

The spectral function acquires the same value for the $2^4 = 16$ different momenta

$$p_\mu + (\pi_H)_\mu \ , \qquad (\pi_H)_\mu = \begin{cases} \pi/b & \text{if } \mu \in H \\ 0 & \text{if } \mu \notin H \end{cases} \tag{3.5}$$

where H is any subset of $\{1, \ldots, 4\}$:

$$\Delta^{-1}(p) = \Delta^{-1}(p + \pi_H). \tag{3.6}$$

This is the spectrum degeneracy problem of the naive lattice Dirac equation: it described 16 times as many fermions as the Dirac equation in the continuum. In contrast to this, there is no spectrum problem for the Kähler equation. The numbers of the fermionic degrees of freedom are the same in the continuum and on the lattice.

L. Susskind [9] has shown in the Hamiltonian formulation how it is possible to enlarge the effective lattice constant by thinning the degrees of freedom, such that the effective 1. Brillouin zone becomes smaller, the degeneracy less dramatic. Sharatchandra, Thun and Weisz have extended this work to the euclidean hypercubic lattice [22]. The result of this reduction is the lattice Kähler equation. For the Hamiltonian approach this was first demonstrated by Dhar and Shankar [23] and in a general coordinate independent form this was done in ref. [17].

b) The spectrum degeneracy group

The reason for the spectrum degeneracy of the naive lattice Dirac equation is a symmetry of this equation described by the spectrum degeneracy group. This group was introduced by Chodos and Healy [12] and discussed f.i. by Sharatchandra, Thun and Weisz [22]. If $\psi(y)$ is a solution of the naive Dirac equation then this is also true for

$$(\hat{M}_H \psi)(y) = e^{iy\pi_H} \ M_H \ \psi(y) \tag{3.7}$$

with $M_\mu := i\gamma_5\gamma_\mu$ and $M_H = M_{\mu_1} \cdot \ldots \cdot M_{\mu_h}$ for $H = (\mu_1, \ldots, \mu_h)$.

Because of the phase factor $\exp(iy \cdot \pi_H)$ in equ. (3.7) the symmetry transformation shifts the momentum of the solution by π_H / b and produces in this way the energy momentum degeneracy. The aim is to decompose the naive lattice Dirac equation by use of the symmetry transformations (3.7). This can be done in two steps:

c) The transition to a new fibering

The transformations $\hat{M}_H$ commute with lattice translations by a multiple of 2 lattice constants. This suggests to consider the lattice of blocks with points x :

$$y = x + \varepsilon_H \,, \quad x^\mu = 2b\, n^\mu \,, \quad n^\mu \in Z \,, \tag{3.8}$$

$a = 2b \equiv$ lattice constant of the block lattice. The index set H measures the deviation of a point in the original y-lattice from the point x of the block lattice (see Fig. 2).

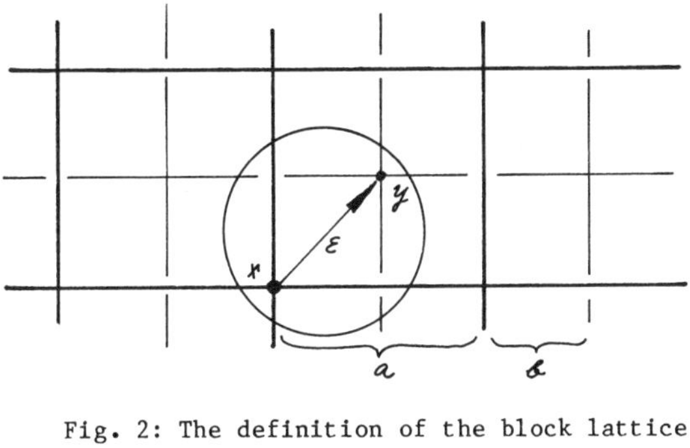

Fig. 2: The definition of the block lattice

The essential step is the reinterpretation of the spinor field $\psi(y)$ which is now considered as a field defined on the points x of the block lattice with additional 16 components which are distributed over the 16 edges of a 4 dimensional cube of side length b at x :

$$\psi(y) = \psi(x + \varepsilon_H) \equiv \psi(x, H). \tag{3.9}$$

Geometrically this means that we pass from a fibre space with the y-lattice as base space and the 4 dimensional spinor space $\mathbb{C}_S^4$ as fibres to a fibre space where the x-block lattice forms the base and the ψ-field takes its values in fibres which are tensor products of the 4 dimensional spinor space with a 16 dimensional space $\mathbb{C}_H^{16}$ corresponding to the index H (see Fig. 3).

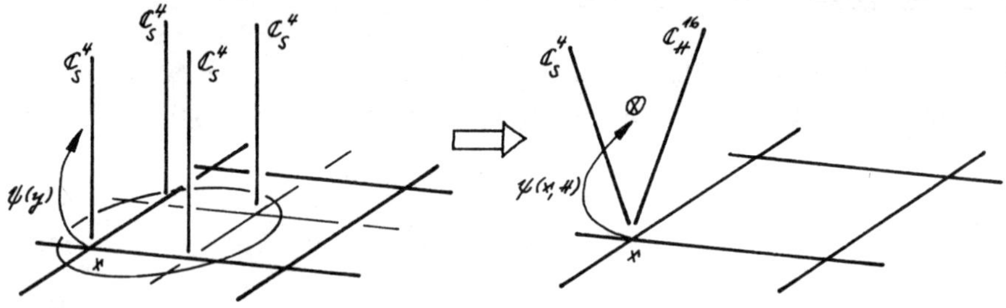

Fig. 3: The geometric interpretation of equ. (3.9)

On the x-lattice, the antisymmetric difference approximation of the naive lattice Dirac equation can be written with the help of block derivatives which act on the x coordinate:

$$
\begin{aligned}
(\nabla_\mu \psi)(x, H) &= (\nabla_\mu^+ \psi)(x, H \setminus \{\mu\}) \quad \text{if } \mu \in H, \\
&= (\nabla_\mu^- \psi)(x, H \cup \{\mu\}) \quad \text{if } \mu \notin H.
\end{aligned}
\tag{3.10}
$$

The Dirac equation then reads

$$
\sum_{\mu \in H} \gamma^\mu \nabla_\mu^+ \psi(x, H\setminus\{\mu\}) + \sum_{\mu \notin H} \gamma^\mu \nabla_\mu^- \psi(x, H\cup\{\mu\}) + m\,\psi(x,H) = 0 \tag{3.11}
$$

and the spectrum degeneracy group acts according to

$$
\hat{M}_H \,\psi(x,K) = e^{i e_K \cdot \pi_H} \, M_H \,\psi(x,K). \tag{3.12}
$$

This is a transformation only in the fibres $\mathbb{C}_S^4 \otimes \mathbb{C}_H^{16}$ but non-trivial on $\mathbb{C}_S^4$ <u>and</u> on $\mathbb{C}_H^{16}$.

d) The spin diagonalization

In the second step we absorb the H-dependent phase factor $\exp(i e_K \cdot \pi_H)$ by a unitary transformation; this is possible since $\hat{M}_H$ and M_H generate the same algebra, the Dirac algebra:

$$\{\hat{M}_\mu, \hat{M}_\nu\} = 2\delta_{\mu\nu} \quad , \quad \{M_\mu, M_\nu\} = 2\delta_{\mu\nu} . \tag{3.13}$$

The transformation is given by

$$\varphi(x,H) = (\gamma_H)^+ \psi(x,H) \quad , \quad \psi(x,H) = \gamma_H \varphi(x,H). \tag{3.14}$$

The degeneracy transformations are accordingly

$$\gamma_H \hat{M} (\gamma_H)^+ = M \tag{3.15}$$

and since M generates the Dirac algebra and acts only on $\mathbb{C}^4_S$ it follows from Schur's lemma that the Dirac operator has to be a multiple of $\mathbb{1}_4$ on $\mathbb{C}^4_S$ (see Fig. 4). This step is called the spin diagonalization [24].

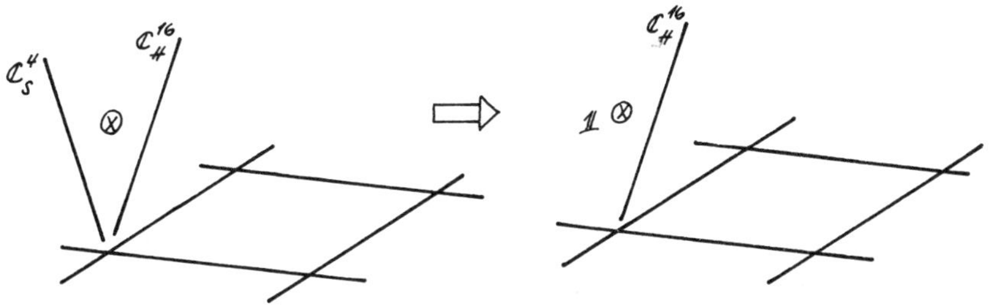

Fig. 4: The geometric interpretation of the spin diagolization step (equ. (3.14)).

The final result is that $\psi(y)$ is a solution of the naive lattice Dirac equation (3.1) if and only if for each $i = 1, \ldots, 4$

$$\varphi_i(x,H) = \sum_a (\gamma_H)^+_{ia} \, \psi_a(x,H) \tag{3.16}$$

is a solution of

$$\sum_{\mu\in H} \rho_{\{\mu\},H\setminus\{\mu\}} \nabla^+_\mu \varphi_i(x,H\setminus\{\mu\}) + \sum_{\mu\notin H} \rho_{\{\mu\},H} \nabla^-_\mu \varphi_i(x,H\cup\{\mu\}) + m\,\varphi_i(x,H) = 0 \tag{3.17}$$

which is the lattice Kähler equation in cartesian components (2.15). The naive Dirac equation decomposes into four Kähler equations. This shows the equivalence of the free Kähler equation on the lattice and the Susskind reduction on the naive Dirac equation. In this way one also gains the continuum limit of Susskind lattice fermions: their limits are Kähler fields.

e) One-component formulation of Susskind fermions

It is possible to reverse the order of the two steps described above. The spin diagonalization of the naive Dirac equation (3.1) leads to a one-component equation (see Fig. 5).

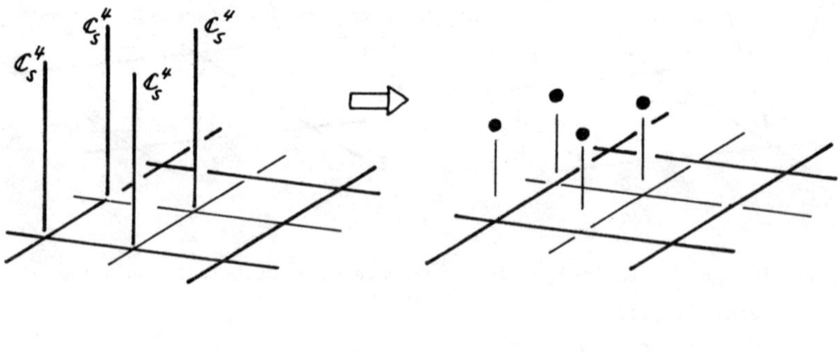

Fig. 5: The spin diagonalization of the naive Dirac equation

Explicitly, one has to find a unitary transformation $T(y) \in \mathcal{U}(4)$ such that

$$T(y)^{+} \gamma_{\mu} T(y+\varepsilon_{\mu}) = c_{\mu}(y) \tag{3.18}$$

is a unitary diagonal matrix. Then, ψ is a solution of equ. (3.1) if and only if $\chi(y) = T(y)^{+}\psi(y)$ is a solution of

$$c_{\mu}(y)\chi(y+\varepsilon_{\mu}) - c_{\mu}(y-\varepsilon_{\mu})^{+}\chi(y-\varepsilon_{\mu}) + m\chi(y) = 0 \tag{3.19}$$

and the $c_{\mu}(y)$ satisfy a cycle condition :

$$c_{p}(y) \equiv c_{\mu}(y)\, c_{\nu}(y+\varepsilon_{\mu})\, c_{\mu}(y+\varepsilon_{\nu})^{+}\, c_{\nu}(y)^{+} = -\mathbb{1}_{4} \,. \tag{3.20}$$

This was shown in ref. [25]. Since the $c_{\mu}(y)$ are diagonal, equ. (3.19) decomposes into 4 one-component equations. The choice of the γ_{μ}-diagonalizing matrices $T(y)$ is not unique. One example is [24,25]

$$T(y) = \gamma_1^{n_1}\gamma_2^{n_2}\gamma_3^{n_3}\gamma_4^{n_4} \quad \text{for } y^{\mu} = b\cdot n^{\mu}. \tag{3.21}$$

It follows from equ. (3.18):

$$c_{\mu}(y) = (-1)^{n_1+\dots+n_{\mu-1}} \,. \tag{3.22}$$

Other possibilities are described in the literature [22,26]. In the Hamiltonian formulation the one-component equation was discussed by Susskind et al. [9].

The one component equation is very convenient for both analytic and numerical calculations on the lattice. On the other hand, the interpretation of the results in terms of flavoured quarks, relevant to the continuum limit is somewhat obscure in this representation.

f) The construction of the fermion field from the one-component field

According to our description of the decomposition of the naive Dirac equation into 4 Kähler equations in subsections c) and d) it is possible to construct a fermion field from the one-component field $\chi(y)$ by a refibering of the fermion space with the help of the block lattice (3.8) (see Fig. 6).

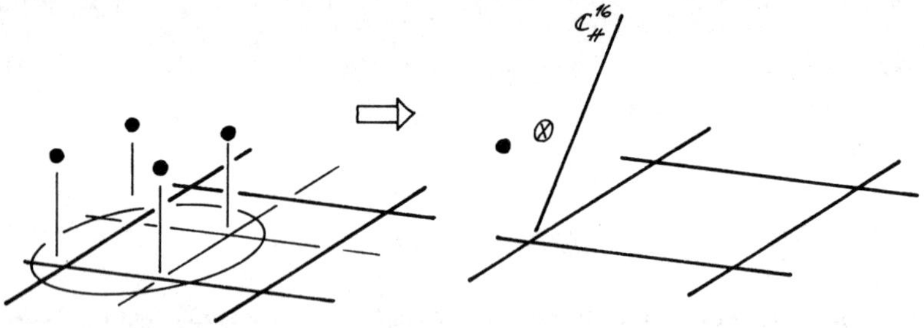

Fig. 6: The geometric interpretation of the transition from the one component field to the Kähler field

Explicitly, it follows from equ. (3.19) [25], that

$$\psi_a^{(b)}(x) = \frac{1}{8} \sum_{H} (\gamma_H)_a{}^b \, \chi(x, H) \tag{3.23}$$

with

$$\chi(x, H) = (-1)^{n_1 + \ldots + n_4} \chi(y) \, , \quad y^\mu = 2b \cdot n^\mu + \varepsilon_H{}^\mu \tag{3.24}$$

satisfies the equation

$$(\Gamma^\mu \nabla_\mu + \tilde{\Gamma}^\mu \tilde{\nabla}_\mu + m)\psi = 0, \tag{3.25}$$

where

$$\{\Gamma^\mu, \Gamma^\nu\} = 2\delta^{\mu\nu} , \; \{\tilde{\Gamma}^\mu, \tilde{\Gamma}^\nu\} = -2\delta^{\mu\nu} , \; \{\Gamma^\mu, \tilde{\Gamma}^\nu\} = 0 \tag{3.26}$$

and

$$\nabla_\mu \psi(x) = \frac{1}{2a}\,(\psi(x+e_\mu) - \psi(x-e_\mu)),$$
$$\tilde{\nabla}_\mu \psi(x) = \frac{1}{2a}\,(\psi(x+e_\mu) + \psi(x-e_\mu) - 2\,\psi(x)). \tag{3.27}$$

The result is again the Kähler equation, because equ.'s (3.25,26) are identical with equ.'s (2.18,19), if we write $\Gamma^\mu = A_+^\mu + A_-^\mu$ and $\tilde{\Gamma}^\mu = A_+^\mu - A_-^\mu$. The reduction procedure leads to special representations of the anticommutation relations (3.26). For instance, one possible representation is

$$\Gamma^\mu = \gamma^\mu \otimes 1_4 \quad , \quad \tilde{\Gamma}^\mu = \gamma_5 \otimes i\gamma^\mu , \tag{3.28}$$

such that the first (second) Dirac matrix acts on the lower (upper) index of the field $\psi_a^{(b)}(x)$. But all representations of equ. (3.26) are equivalent [18].

4) The Kähler field interacting with a gauge field

a) The gauge coupling in the continuum

The coupling of the Kähler field to an external gauge potential which acts on an additional colour index $\phi_c = \sum_H \varphi_c(x,H)\, dx^H$ is straightforward in

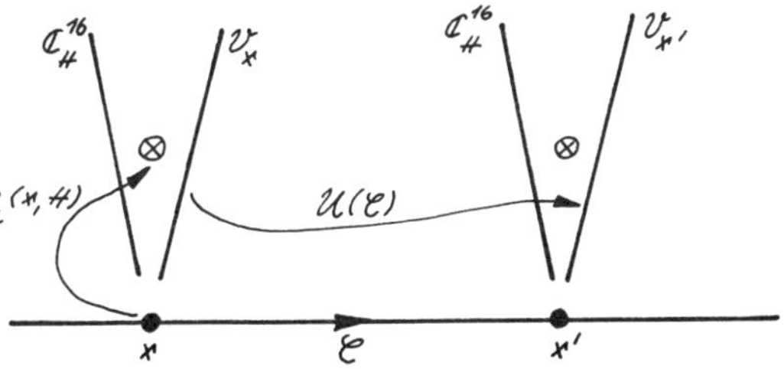

Fig. 7: The geometric interpretation of the Kähler field with gauge interaction

the continuum. Geometrically, we introduce local colour spaces V_x and path dependent parallel transporters $\mathcal{U}(\mathcal{C})$ between colour spaces at different points $x \xrightarrow{\mathcal{C}} x'$ (see Fig. 7).

Analytically, this amounts to a replacement of the derivatives by covariant derivatives. By this, the Kähler equation (2.31) becomes

$$(d_A - \delta_A + m)\,\phi \equiv (dx^\mu \vee \mathcal{D}_\mu + m)\,\phi = 0 \tag{4.1}$$

where

$$\begin{aligned}(\mathcal{D}_\mu \phi)_c(x,H) &= \partial_\mu \phi_c(x,H) - i\,(\mathcal{A}_\mu(x))_c{}^{c'}\,\phi_{c'}(x,H),\\ \mathcal{A}_\mu(x) &= g\,\frac{\lambda_a}{2}\,A_\mu^a(x),\end{aligned} \tag{4.2}$$

is the covariant derivative and d_A and δ_A are the covariant exterior derivative and the covariant generalized divergence, respectively:

$$d_A\,\phi = dx^\mu \wedge \mathcal{D}_\mu \phi\,, \qquad \delta_A\,\phi = -e^\mu \lrcorner\, \mathcal{D}_\mu \phi$$

(see equ.'s (2.3,26)). From the identity

$$dx^\mu \vee \mathcal{D}_\mu = dx^\mu \vee \partial_\mu - i\,\mathcal{A} \vee\,, \qquad \mathcal{A} = \mathcal{A}_\mu\, dx^\mu \tag{4.3}$$

it follows that equ. (4.1) can be written in the form

$$(d - \delta + m)\,\phi = i\mathcal{A} \vee \phi\,. \tag{4.4}$$

Therefore, the gauge symmetry commutes with the projection to the flavour components (equ. (2.37)). Equ. (4.4) is again equivalent to 4 Dirac equations with gauge interaction.

b) The gauge coupling on the lattice

The coupling of the lattice Kähler field to an external gauge potential is not unique. The free Kähler equation is equivalent to Susskind's formulation of lattice fermions. Therefore it is possible to couple the Kähler field accordingly. Even in this case, the geometrically intuitive formulation of fermions in the Kähler framework is superior to other methods f.i. in connection with the path integral quantization of the theory or with the discussion of symmetries and currents (see subsections c)-f) below). However, there are alternatives to Susskind's way of gauging and it is worth-while to have a look at these possibilities. Unfortunately, there is up to now only one definite suggestion of a gauge coupling of the lattice Kähler field in the literature [17] which is inequivalent to Susskind's method. It consists of a literal translation of the continuum procedure: the gauged lattice Kähler equation is given by

$$(\breve{\Delta}_A - \breve{\nabla}_A + m)\phi = 0 \tag{4.5}$$

(compare equ. (4.1)), where $\breve{\Delta}_A$ and $\breve{\nabla}_A$ are the covariant dual boundary and covariant dual coboundary operator, respectively. With the help of the covariant difference operators

$$\begin{aligned}(\mathcal{D}_\mu^+ \varphi)(x, H) &= \frac{1}{a}\left(\mathcal{U}(x,\mu)\, \varphi(x+e_\mu, H) - \varphi(x, H)\right),\\ (\mathcal{D}_\mu^- \varphi)(x, H) &= \frac{1}{a}\left(\varphi(x, H) - \mathcal{U}(x-e_\mu,\mu)^{-1}\, \varphi(x-e_\mu, H)\right),\end{aligned} \tag{4.6}$$

they are defined by

$$\begin{aligned}\breve{\Delta}_A \phi &= \sum_{x,H}\Big(\sum_{\mu\in H} \rho_{\{\mu\}, H\setminus\{\mu\}}\, (\mathcal{D}_\mu^+ \varphi)(x, H\setminus\{\mu\})\Big)\, d^{x,H},\\ \breve{\nabla}_A \phi &= \sum_{x,H}\Big(\sum_{\mu\notin H} \rho_{\{\mu\}, H}\, (\mathcal{D}_\mu^- \varphi)(x, H\cup\{\mu\})\Big)\, d^{x,H}\end{aligned} \tag{4.7}$$

in cartesian coordinates. $\mathcal{U}(x,\mu)$ maps the local colour space $\mathcal{V}_{x+e_\mu}^t$ onto $\mathcal{V}_x^t$.

The equations of motion (4.5) for the Kähler field ϕ and its (in euclidean space) independent adjoint field $\bar{\phi}$ can be derived from the action

$$S = \frac{1}{4}\left(\bar{\phi}, (\breve{\Delta}_A - \breve{\nabla}_A + m)\phi\right)_0 (v), \tag{4.8}$$

where $\mathcal{V} = \sum_x (x,\{1,2,3,4\})$ is the volume chain and $(\Xi,\Phi)_0$ is the (zeroth) scalar product between co-chains defined by

$$(\Xi,\Phi)_0 = \sum_x \Big(\sum_H \xi(x,H)\, \varphi(x,H)\Big)\, d^{\,x,\{1,2,3,4\}} \tag{4.9}$$

in cartesian coordinates. The scalar product (4.9) is a 4 co-chain. The operation which corresponds to integration in the contiuum is the application to the volume chain.

For a comparison with other approaches, I rewrite the action (4.8) in Dirac components (2.28-32):

$$\begin{aligned} S = a^4 \sum_{x,\mu} \{ \tfrac{1}{a} \big(& \bar\psi(x) A_+^\mu\, \mathcal{U}(x,\mu)\, \psi(x+e_\mu) - \bar\psi(x+e_\mu)\, \mathcal{U}(x,\mu)^{-1} A_-^\mu\, \psi(x) \\ & - \bar\psi(x) A_+^\mu\, \psi(x) \qquad + \bar\psi(x) A_-^\mu\, \psi(x)\big) \} \\ + a^4 \sum_x \ & m\, \bar\psi(x)\, \psi(x) \end{aligned} \tag{4.10}$$

$$\begin{aligned} = a^4 \sum_{x,\mu} \{ & \tfrac{1}{2a} \big(\bar\psi(x)\, \Gamma^\mu\, \mathcal{U}(x,\mu)\, \psi(x+e_\mu) - \bar\psi(x+e_\mu)\, \mathcal{U}(x,\mu)^{-1}\, \Gamma^\mu\, \psi(x)\big) \\ & + \tfrac{1}{2a} \big(\bar\psi(x)\, \tilde\Gamma^\mu\, \mathcal{U}(x,\mu)\, \psi(x+e_\mu) + \bar\psi(x+e_\mu)\, \mathcal{U}(x,\mu)^{-1}\, \tilde\Gamma^\mu\, \psi(x) \\ & \qquad - 2\bar\psi(x)\, \tilde\Gamma^\mu\, \psi(x)\big) \} \\ + a^4 \sum_x \ & m\, \bar\psi(x)\, \psi(x). \end{aligned}$$

$A_\pm^\mu$ and $\Gamma^\mu, \tilde\Gamma^\mu$ are the hypercomplex numbers introduced in equ.'s (2.19) and (3.26), respectively. The formal continuum limit of the action (4.10) is given by

$$\begin{aligned} S = a^4 \sum_x \{ & \bar\psi(x)\, \Gamma^\mu D_\mu\, \psi(x) + m\, \bar\psi(x)\, \psi(x) \\ & + \tfrac{a}{2}\, \bar\psi(x)\, \tilde\Gamma^\mu D_\mu^2\, \psi(x) + O(a^2) \} \end{aligned} \tag{4.11}$$

up to terms of the order a^2. It is inequivalent to what one derives from Susskind's one-component equation. In that case, one introduces $2^4 = 16$ times as many link variables as in the Kähler equation (4.5), because it is natural in that formulation to attach one colour space $\mathcal{V}_y$ to each point of the sublattice with half the lattice constant compared to the lattice constant of the Kähler formulation. The result is [25] that the naive continuum limit of the gauge invariant Susskind action shows an additional $O(a)$ term compared to equ. (4.11) which is proportional to the field strength tensor $\mathcal{F}_{\mu\nu}$. Such a term cannot emerge in the formulation (4.5,8) because the action involves only nearest neighbour terms. This is different in the one-component approach (3.19) because the components χ must be transported along paths which are up to 4 sublattice units long, in order to construct the gauge covariant quark field (see equ.'s (3.23,24)) [25]. The quark field is an extended object in the one-component formulation [26].

There is a serious objection to the gauge coupling (4.5-8) of the Kähler field. It has been shown in weak coupling perturbation theory [27] as well as at $g^2 = \infty$ and in leading order of $1/N \to 0$ [28] that an unwanted chiral invariant counter term is induced in the quantized theory which is forbidden in the one-component approach. The reason for this is that the action (4.8) is less symmetric than Susskind's one. This suggests to use the latter or to look for different coupling schemes, f.i. with a more symmetric interpretation of the gauge field on the lattice or with the real Kähler equation as a starting point.

<u>c) The quantization of the Kähler field</u>

Given the action, f.i. the gauge invariant action (4.8) of the Kähler field, the path integral quantization is straightforward. For instance, it is possible to calculate the usual 2n-point Green functions $\langle \phi \ldots \phi \bar{\phi} \ldots \bar{\phi} \rangle$, which are multi-differential forms [29] and multi-cochains in the continuum and on the lattice, respectively. As an example, the free propagator necessary for the weak coupling perturbation expansion is given in Dirac components by

$$\langle \psi_a^{(b)}(x) \bar{\psi}_{a'}^{(b')}(x') \rangle = \left((A_+^\mu)_{aa'}^{bb'} \nabla_\mu^+ + (A_-^\mu)_{aa'}^{bb'} \nabla_\mu^- - m \delta_{aa'} \delta^{bb'} \right) \Delta_S(x-x'). \quad (4.12)$$

$\Delta_S(x)$ is the propagator of the free scalar field

$$\Delta_S(x) = \frac{-1}{(2\pi a)^4} \int_{-\pi}^{\pi} d^4\beta \, \frac{e^{-i\beta x/a}}{\hat{q}_0(\beta) + m^2}, \quad \hat{q}_0(\beta) := \sum_\mu \left(\frac{2}{a} \sin\frac{\beta_\mu}{2}\right)^2. \tag{4.13}$$

Using Schwinger's proper time method [30] Δ_S can be evaluated explicitly:

$$\Delta_S(x) = \frac{1}{i a^4} \int_0^\infty ds \, e^{-i(m^2 + \frac{8}{a^2})s} \prod_{\mu=1}^{4} I_{\frac{|x^\mu|}{a}}\left(\frac{2is}{a^2}\right). \tag{4.14}$$

The parameter integral is Lauricella's generalization of the hypergeometric function to several variables [31]. Equ. (4.12) shows that the computation methods with quantized lattice Kähler fields are completely analogous to the continuum case.

It is also possible to calculate the effective fermion action. For this, we write the action as a bilinear form in the fields:

$$S = a^4 \sum_{x_1, x_2} \bar{\psi}(x_1) \, Q_{x_1 x_2}(\mathcal{U}) \, \psi(x_2) \tag{4.15}$$

with f.i.

$$\begin{aligned} Q_{x_1 x_2}(\mathcal{U}) = {} & \Big(m - \sum_{\pm\mu} A_+^\mu\Big) \frac{1}{a^4} \delta_{x_1, x_2} + \\ & + a^4 \sum_{x, \pm\mu} \frac{1}{a} A_+^\mu \, \mathcal{U}(x,\mu) \frac{1}{a^4} \delta_{x_1, x} \frac{1}{a^4} \delta_{x+e_\mu, x_2}, \end{aligned} \tag{4.16}$$

$$A_-^\mu = -A_+^{-\mu}, \quad \mathcal{U}(x - e_\mu, \mu)^{-1} = \mathcal{U}(x, -\mu), \quad e_{-\mu} = -e_\mu,$$

for the action (4.8). Then, the effective fermion action is given by

$$S_{eff} = \log \det Q(\mathcal{U}) \doteq \operatorname{trace} \log(1 - \kappa M) \tag{4.17}$$

with $\kappa = (ma + \sum_{\pm\mu} A_+^\mu)/(m^2 a^2 + 4)$, $M = \sum_{\pm\mu} M_\mu$, $M_\mu = -A_+^\mu \, \mathcal{U}(x,\mu)$.

This is the starting point of the hopping parameter expansion for Kähler fermions. However, Monte Carlo calculations based on the Kähler effective action (4.17) have not been done, yet.

These two examples should indicate that one expects no conceptional problems with the path integral quantization of Kähler fields.

d) Symmetries and currents

The clear advantage of the Kähler formulation is its close connection with the continuum theory. This allows the translation of physically relevant quantities to the lattice in an algorithmic way. I would like to illustrate this procedure by way of an example, which has even led to controversies in the literature for the one-component formulation [32], the definition and interpretation of symmetries and currents.

The starting point is the following Green's formula for the Kähler operator

$$\check{\Delta}(\bar{\phi}, \Sigma)_1 = (\bar{\phi}, (\check{\Delta} - \check{\nabla})\Sigma)_0 + ((\check{\Delta} - \check{\nabla})\bar{\phi}, \Sigma)_0 \tag{4.18}$$

where $(\bar{\phi}, \Sigma)_0$ is the zeroth scalar product (4.9) between co-chains and

$$(\phi, \Sigma)_1 := \sum_p (-1)^{\binom{p}{2}} \, {}^p\phi \wedge \star\, {}^{p-1}\Sigma + \phi \curlywedge \Sigma \tag{4.19}$$

is the so-called first derived scalar product, which is a 3-co-chain; it follows that the 1-co-chain

$$j = \star^{-1}(\bar{\phi}, \Sigma)_1 = \sum_{x,\mu} j_\mu(x)\, d^{x,\{\mu\}} \tag{4.20}$$

is a conserved current

$$\check{\nabla} j = (-\underbrace{\star^{-1}\check{\Delta}\star)(\star^{-1}}_{=1}(\bar{\phi}, \Sigma)_1) = 0 \tag{4.21}$$

if Ξ and $\bar{\Phi}$ are solutions of the free Kähler equation and the adjoint free Kähler equation, respectively.

Because the definition of the first derived scalar product involves the exterior product, the matching condition (equ. (2.42) and Fig. 1) for this product introduces a nearest neighbour point splitting. In the Dirac basis, for instance, $j_\mu(x)$ is given by

$$j^\mu(x) = \bar{\varphi}(x)\, A_+^\mu\, \xi(x+e_\mu) + \bar{\varphi}(x+e_\mu)\, A_-^\mu\, \xi(x). \tag{4.22}$$

In order to make this current gauge invariant in a theory with gauge interaction, we insert parallel transport matrices

$$j^\mu(x) = \bar{\varphi}(x)\, A_+^\mu\, \mathcal{U}(x,\mu)\, \xi(x+e_\mu) + \bar{\varphi}(x+e_\mu)\, \mathcal{U}(x,\mu)^{-1} A_-^\mu\, \xi(x). \tag{4.23}$$

This current may be viewed as the natural gauge invariant point split lattice current for quantized Kähler fields. It is the result of the tight lattice-continuum analogy for geometric objects. I give examples:

e) The vector current

From the reality of the Kähler operator it follows that the action is invariant with respect to global continuous phase transformations of the complex Kähler field:

$$\Phi \mapsto e^{i\alpha}\Phi\,, \qquad \bar{\Phi} \mapsto \bar{\Phi}\, e^{-i\alpha}. \tag{4.24}$$

In the Dirac basis, the generator of these transformations is given by $\mathbb{1}_4 \otimes \mathbb{1}_4$ (see equ. (3.28)). The conserved current connected with this symmetry is

$$j = i\,\mathcal{A}^{-1}(\bar{\Phi},\Phi)_1\,, \qquad \check{\nabla} j = 0. \tag{4.25}$$

In the Dirac basis and in the presence of an external gauge field this becomes

$$j^{\mu}(x) = i\left(\bar{\psi}(x) A_{+}^{\mu} \mathcal{U}(x,\mu)\psi(x+e_{\mu}) + \bar{\psi}(x+e_{\mu})\mathcal{U}(x,\mu)^{-1} A_{-}^{\mu}\psi(x)\right). \tag{4.26}$$

Again, this current is conserved:

$$\nabla_{-}^{\mu} j_{\mu}(x) = 0.$$

The naive continuum limit of the vector current (4.26) is

$$i\bar{\psi}\,\Gamma^{\mu}\psi = i\sum_{b}\bar{\psi}^{(b)}\gamma^{\mu}\psi^{(b)}. \tag{4.27}$$

f) Chiral currents

There are two transformations of the free continuum Kähler equation which anticommute with the Kähler operator $d-\delta$ and are therefore symmetries only for $m=0$. The first of these symmetries is the flavour neutral chiral symmetry $\delta\phi = \varepsilon \vee \phi\, \delta\beta$ which is in the Dirac basis given by $(\gamma_5 \otimes 1\!\!1_4)\psi\, \delta\beta$. The second one is the symmetry $\delta\phi = (\varepsilon \vee \phi \vee \varepsilon)\delta\beta$ which is a simultaneous transformation in Dirac space and in flavour space described by $(\gamma_5 \otimes \gamma_5)\psi\, \delta\beta$ in the Dirac basis. For completeness, I mention a third symmetry, namely $\delta\phi = (\phi \vee \varepsilon)\, \delta\beta$, which is a pure vector symmetry, generating a subgroup of the flavour symmetry group (2.35,36). The second and third of these transformations are called Euler and Hirzebruch transformations according to a paper by Nielsen, Römer and Schroer [33] where the anomaly of the associated currents was calculated in curved space time.

The chiral symmetry on the lattice can be formulated with the help of the lattice analogue of the volume form, the 4-co-chain $\varepsilon = \sum_{x} d^{x,\{1,2,3,4\}}$. $\phi \mapsto \varepsilon \vee \phi$ is a symmetry of the free massless Kähler equation:

$$(\check{\Delta} - \check{\nabla})(\varepsilon \vee \phi) = -\varepsilon \vee (\check{\Delta} - \check{\nabla})\phi. \tag{4.28}$$

Because $\exp(i\frac{\pi}{2}\varepsilon\vee) = i\varepsilon\vee$ in the continuum, $\phi \mapsto \varepsilon\vee\phi$ is a finite chiral rotation. The free massless Kähler equation has a discrete chiral invariance in agreement with its equivalence with the Susskind equation. In accordance with the existing no-go theorems of Nielsen and Ninomiya [34] this symmetry cannot be extended to a continuous local symmetry on the lattice. In the geometric formulation the reason for this is the matching condition for the Clifford product $\vee$ which makes the action of $\varepsilon\vee$ local but not point-like. Therefore, $\exp(i\beta\varepsilon\vee)$ is a non-local operator on the lattice.

A reasoning for the non-pointlike action of $\varepsilon\vee$ was also given by Rabin [19]: his argument is that $\varepsilon\vee$ is essentially the $\star$-operation: $\varepsilon\vee {}^p\phi = (-1)^p \star {}^p\phi$. ϕ changes to $\phi+\delta\phi$ by an infinitesimal transformation, but $\phi+\delta\phi$ is defined only, if both ϕ and $\delta\phi$ live on the same lattice. This is not true for $\delta\phi = \star\phi\,\delta\beta$ because $\star$ transforms to the dual lattice. One has to define $\delta\phi$ with an extra isomorphism between the lattice and its dual. This, however, involves a translation.

Nevertheless, the discrete chiral symmetry leads to conserved currents

$$j_5 = \frac{i}{2}\star^{-1}(\bar{\phi},\, \varepsilon\vee\phi)_1 + h.c. \;=\; \sum_{x,\mu} j_{5,\mu}(x)\, d^{x,\{\mu\}}$$

and (4.29)

$$k_5 = \frac{i}{2}\star^{-1}(\bar{\phi},\, \varepsilon^{-1}\vee\phi)_1 + h.c. \;=\; \sum_{x,\mu} k_{5,\mu}(x)\, d^{x,\{\mu\}},$$

which are different because $\varepsilon\vee$ is not an involution on the lattice. For completeness, I would like to mention that there are two additional currents on the lattice which are conserved for $m=0$. These are of the order of the lattice constant but they control fluctuations in the quantized lattice theory. The following discussion is restricted to j_5.

In Dirac components, j_5 is given by

$$j_5^{\mu}(x) = \frac{i}{2}\left(\bar{\psi}(x)A_+^{\mu}(\Gamma_5\psi)(x+e_\mu) + \bar{\psi}(x+e_\mu)A_-^{\mu}(\Gamma_5\psi)(x) - h.c.\right) \qquad (4.30)$$

with

$$(\Gamma_5 \psi)(x) = \sum_{s_1,\ldots,s_4=\pm 1} A^1_{s_1} A^2_{s_2} A^3_{s_3} A^4_{s_4} \psi(x+e_H), \quad H=\{\mu/s_\mu=+1\}. \tag{4.31}$$

This shows explicitly that the geometric lattice description of currents introduces an additional point splitting. The divergence of the current (4.30) is zero for massless Kähler fields:

$$\nabla^\mu_- j_{5\mu}(x) = 2im \cdot \frac{1}{2}(\bar{\psi}(x)(\Gamma_5\psi)(x) + h.c.). \tag{4.32}$$

It is also easy to check that $\Gamma_5 = \gamma_5 \otimes 1\!\!1_4 + O(a)$ in the naive continuum limit.

For a theory with gauge interaction we have to generalize the definition of the Γ_5 -operator such that $(\Gamma_5\psi)(x)$ takes its values in the local colour spaces $\mathcal{V}_x$; for this, we substitute the translations in the definition (4.31) of Γ_5 by covariant translations:

$$(\Gamma_5 \psi)(x) = \sum_{s_1,\ldots,s_4=\pm 1} A^1_{s_1} \cdots A^4_{s_4} \; \mathcal{U}(x,H)\, \psi(x+e_H) \tag{4.33}$$

where

$$\mathcal{U}(x,H) = \mathcal{U}(x,1)^{x_1} \cdot \mathcal{U}(x+e_1,2)^{x_2} \cdot \mathcal{U}(x+e_1+e_2,3)^{x_3} \cdot \mathcal{U}(x+e_1+e_2+e_3,4)^{x_4} \tag{4.34}$$

for $e_H = a\cdot(x_1,\ldots,x_4)$. Now it turns out that j_5 is no longer conserved but that for the action (4.8) the divergence converges to the correct anomaly in the continuum limit [35]:

$$\begin{aligned}
\nabla^\mu_- j_{5,\mu} &= M + A, \\
M(x) &\xrightarrow{a\to 0} 2mi \sum_{G} \bar{\psi}^{(G)}(x)\, \gamma_5\, \psi^{(G)}(x), \\
A(x) &\xrightarrow{a\to 0} \underbrace{4}_{N_F} \cdot (g^2/16\pi^2 i)\; \mathrm{trace}_c\, \mathcal{F}^{\mu\nu}(x)\, \varepsilon_{\mu\nu\rho\sigma}\, \mathcal{F}^{\rho\sigma}(x).
\end{aligned} \tag{4.35}$$

In this way, the point splitting of the currents which is an intrinsic property of the geometric lattice approach turns out to be one correct field theoretic point splitting [36].

The Euler symmetry on the lattice is a global continuous symmetry. The decomposition

$$e^{i\alpha(\gamma_5\otimes\gamma_5)} = e^{i\alpha\gamma_5}\otimes\frac{1+\gamma_5}{2} + e^{-i\alpha\gamma_5}\otimes\frac{1-\gamma_5}{2} \tag{4.36}$$

shows that there are two flavours transforming with a positive chiral charge and two flavours transforming with a negative chiral charge. Therefore, this continuous symmetry is again not in conflict with the no-go theorems [34], the numbers of left- and right-handed species are equal. A consequence of this is that the associated current

$$\begin{aligned} j_\epsilon &= i\,\bigstar^{-1}(\bar{\Phi}, \epsilon \vee \Phi \vee \epsilon)_1 \ , \quad \text{i.e.} \\ j_\epsilon^{\,\mu}(x) &= i\,(\bar{\psi}(x) A_+^\mu\, \mathcal{U}(x,\mu)\, \gamma_5\otimes\gamma_5\, \psi(x+e_\mu) + \\ &\quad + \bar{\psi}(x+e_\mu)\, \mathcal{U}(x,\mu)^{-1}\, \gamma_5\otimes\gamma_5\, A_-^\mu\, \psi(x)) \end{aligned} \tag{4.37}$$

is anomaly free:

$$\nabla_-^\mu\, j_{\epsilon,\mu}(x) = 2im\, \bar{\psi}(x)\, \gamma_5\otimes\gamma_5\, \psi(x). \tag{4.38}$$

The Euler symmetry was shown to be spontaneously broken in the strong coupling approximation [9,37]. It is of course possible to study other symmetries f.i. those connected with the flavour symmetry group [17]. This leads to currents $j = \frac{1}{4}\bigstar^{-1}(\bar{\Phi}, \Phi\vee c)$ which are exactly conserved (!) for the free Kähler equation and for every transformation c (see equ. (2.34)).

At the very end of my talk I would like to stress that the Kähler equation leads to a geometrical formulation of the kinematics of fermions on the lattice. Also in this context there are still more open questions than solved problems or worked-out applications. But the real problem of fermions on the lattice is a dynamical one. I am sure, that the investigation of dynamical models f.i. realistic formulations of lattice QCD or supersymmetric lattice models will prove the advantages of the geometrical approach.

References

1. E. Kähler; Rend. Math. Ser. V, 21 (1962) 425.
2. I.M. Benn, R.W. Tucker; Commun. Math. Phys. 89 (1983) 341.
3. W. Graf; Ann. Inst. H. Poincaré, Sect. A 29 (1978) 85.
4. I.M. Benn, R.W. Tucker; Phys. Letters B119 (1982) 348.
5. T. Banks, Y. Dothan, D. Horn; Phys. Letters B117 (1982) 413.
6. B. Holdom; 'Gauged fermions from tensor fields', Stanford Univ. Preprint ITP - 738, 3/83 (1983).
7. I.M. Benn, R.W. Tucker; J. Phys. A 16 (1983) L 123.
8. I.M. Benn, R.W. Tucker; Phys. Letters B 125 (1983) 47 and 'Clifford analysis of exterior forms and Fermi-Bose symmetry', University of Lancaster Preprint (1983).
9. L. Susskind; Phys. Rev. D 16 (1977) 3031.
   T. Banks, S. Raby, L.Susskind, J. Kogut, D.R.T. Jones, P.N. Scharbach, D. Sinclair; Phys. Rev. D 15 (1976) 1111.
10. P. Becher, H. Joos; 'On the geometric lattice approximation to a realistic model of QCD', Preprint DESY 82-088 (1982) to appear in Nuovo Cim. Letters.
11. P. Mitra; Phys. Letters B 123 (1983) 77.
    A.N. Burkitt, A. Kenway, R.D. Kennway; 'A practical scheme for SU(4) flavour symmetry breaking in lattice gauge theories', Edingburgh Preprint 83/249 (1983).
12. A. Chodos, J.B. Healy; Nucl. Phys. B 127 (1977) 426.
13. H. Raszillier; J. Math. Phys. 24 (1983) 642.
    W. Celmaster; Phys. Rev. D 26 (1982) 2955.
    W. Celmaster, F. Krausz; 'Fermion mutilation on a body-centered tesseract', Preprint NUB 2596 Boston (1983).
14. J.M. Drouffe, K.J.M. Moriarty; 'Gauge theories on a simplicial lattice', Preprint Ref. Th. 3518-CERN (1983).
15. M. Göckeler; 'Dirac-Kähler fields and the lattice shape dependence of fermion flavour', Preprint DESY 82-082 (1982).
16. H. Raszillier; 'Lattice degeneracies of geometric fermions', Preprint Bonn-HE-83-7 (1983).
17. P. Becher, H. Joos; Z. Physik C - Particles and Fields 15 (1982) 343.
18. P. Becher; Phys. Letters B 104 (1981) 221.
19. J.M. Rabin; Nucl. Phys. B 201 (1982) 315.
20. K. Wilson; in 'New phenomena in subnuclear physics', A. Zichichi, ed. ("Erice 1975") New York, Plenum 1977.

21. S.D. Drell, M. Weinstein, S.Yankielowicz; Phys. Rev. D 14 (1976) 487, 1627.

22. H.S. Sharatchandra, H.J. Thun, P. Weisz; Nucl. Phys. B 192 (1981) 205.

23. A. Dhar, R. Shankar; Phys. Letters B 113 (1982) 391.

24. N. Kawamoto, J. Smit; Nucl. Phys. B 192 (1981) 100.

25. H. Kluberg-Stern, A. Morel, O. Napoly, B. Petersson; 'Flavours of Lagrangian Susskind fermions', CEN-SACLAY-Preprint DPh. G. SPT /83/29 (1983).

26. F. Gliozzi; Nucl. Phys. B 204 (1982) 419.

27. P. Mitra, P. Weisz; 'On bare and induced masses of Susskind fermions', Preprint DESY 83-013 (1983).

28. O.Napoly; 'Absence of Goldstone boson for a U(N) gauge theory with Dirac-Kähler fermions on the lattice', CEN-SACLAY Preprint SPh. T /83/77 (1983).

29. H.K. Nickerson, D.C. Spencer, N.E. Steenrod; Advanced Calculus (van Nostrand Comp., Princeton, N.J. (1959)).

30. J. Schwinger; Phys. Rev. 82 (1951) 664.

31. A. Erdelyi et al., eds.; Tables of Integral Transforms I (McGraw-Hill, New York (1954)).

32. T. Banks, A. Zaks; Nucl. Phys. B 206 (1982) 23.
C. Rebbi; 'Monte Carlo computations of the hadronic mass spectrum', Proceedings of the 19th Orbis Scientae Meeting 1982.

33. N.K. Nielsen, H. Römer, B. Schroer; Nucl. Phys. B 136 (1978) 475.

34. H.B. Nielsen, M. Ninomiya; Nucl. Phys. B 185 (1981) 20 (Erratum: Nucl. Phys. B 195 (1982) 541); Nucl. Phys. B 193 (1981) 173 and Phys. Letters B 105 (1981) 219.

35. M. Göckeler; 'Axial-vector anomaly for Dirac-Kähler fermions on the lattice', Preprint DESY 83-009 (1983).

36. J. Schwinger; Phys. Rev. Letters 3 (1959) 296.

37. H. Kluberg-Stern, A. Morel, B. Petersson; Phys. Letters B 114 (1982) 152 and Nucl. Phys. in press.

θ PARAMETER MONTE-CARLO*

G. Bhanot

The Institute for Advanced Study
Princeton, NJ 08540, USA

ABSTRACT

I describe recent results from a Monte-Carlo simulation of SU(2) gauge theory in 4 dimensions including the vacuum angle Θ. In the simulation, a new method due to Woit was used to measure the lattice topological charge. This method is also described.

I. Introduction and Method

It is well known that QCD has a hidden parameter, the vacuum angle Θ. For the continuum theory, one defines the topological charge

$$Q = \frac{1}{16\pi^2 N} \int \mathrm{tr}\ \{F_{\mu\nu}(x)\tilde{F}_{\mu\nu}(x)\}d^4x \tag{1}$$

with

$$F_{\mu\nu} = \partial_\mu A_\nu - \partial_\nu A_\mu + [A_\mu, A_\nu] \tag{2}$$

and

$$\tilde{F}_{\mu\nu} = \varepsilon_{\mu\nu\alpha\beta}F_{\alpha\beta}\ . \tag{3}$$

The partition function is then given by

$$Z_{\text{Minkowski}} = \int \pi dA_\mu(x)\ e^{iS_g} \cdot e^{i\Theta Q} \tag{4}$$

---

*Paper presented at the 7th Johns Hopkins Workshop on 'Current Problems in High Energy Physics', 21-23 June, 1983, Bad Honnef, Germany. The work described was done in collaboration with E. Rabinovici, N. Seilberg and P. Woit.

with S_g the usual action ($\sim \int F^2$).

The effects of Q have been used to discuss various aspects of the U(1) problem, possible mechanisms for chiral symmetry breaking, confinement in QCD, the large mass of the η', oblique confinement phases[1,2] etc. It is then of great interest to study these effects in the non-perturbative context of lattice gauge theory.

This was first attempted by Peskin[3] who gave a straightforward definition for Q using a product of eight lattice link elements to define the lattice analogue of $F\tilde{F}$. A slightly different definition, but in the same spirit, was given by diVecchia, Fabricius, Rossi and Veneziano[4].

However, neither of these definitions is completely satisfactory. The quantities used in References 3 and 4 have perturbative pieces from higher order operators while the continuum $F\tilde{F}$ has no such perturbative parts. These higher order operators are not expressible as total derivatives and so, the topological significance of $F\tilde{F}$ is obscured.

I will now describe a method due to Peter Woit[5] at Princeton which seems much more direct in a topolotical sense, is easy to implement in a Monte-Carlo simulation and gives the correct weak coupling behavior (scaling). The rest of this section is essentially a rephrasing of Woit's paper.

The lattice definition is made transparent by the following discussion in the continuum theory:

Since $F\tilde{F}$ is a total divergence, Q is different from zero only if A is "singular". Pick a gauge where this singularity is not at infinity. Now one can remove the singularity locally (say in a region R surrounding it) by doing gauge transformations. The information about the topology of the original singular configuration is now contained in the mapping of the gauge function from the boundary of R to the outside. The value of Q (the degree of mapping) is the number of times the gauge function that removed the singularity wraps around the group manifold before giving the configuration on the outside of R.

We will use this picture to determine Q for any given SU(2) configuration of links at large values of the inverse coupling (i.e. in the continuum limit). First pick a time direction and orthogonal three tori. On each three torus, smooth the space-like links by finding the gauge function $\Phi(\vec{x},t)$ which maximizes

$$\sum_{\substack{i,\vec{x},t \text{ fixed} \\ i \text{ spacelike}}} \text{tr}\,\{\Phi(\vec{x},t)\; U_i(\vec{x},t)\Phi^{-1}(\vec{x}+\hat{i},t)\}\;. \tag{5}$$

This can be achieved by the following iterative procedure: Start with any $\{\Phi(\vec{x},t),\ \vec{x}$ ranges over three torus$\}$ which is not identically zero. Maximize

$$\Sigma_i\ \text{tr}\,\{\Phi(\vec{x},t)\; U_i(\vec{x},t)\Phi^{-1}(\vec{x}+\hat{i},t)\}. \tag{6}$$

This can be done by replacing $\Phi(\vec{x},t)$ with the "average" value of χ with

$$\chi^{-1} = \Sigma_i\ U_i(\vec{x},t)\Phi^{-1}(\vec{x}+\hat{i},t).$$

To compute χ^{-1} one adds the six SU(2) elements viewed as vectors in R^4 and then projects back onto S^3. In practice, this procedure, when iterated about 5 to 10 times per timelike hyper-slice, gives a good determination of $\Phi(\vec{x},t)$. The idea is that $\Phi(\vec{x},t)$ has the effect of putting the links on each time slice into the same gauge.

Now, in analogy with the continuum example mentioned before, we compute the degree of the gauge function Φ between time slices. This is done as follows:

Φ is a function in SU(2) at each site $(\vec{x},t)$.

Define

$$\hat{\Theta}_t(\vec{x}) = \Phi^{-1}(\vec{x},t+1)\; U_0\Phi(\vec{x},t). \tag{7}$$

It is this function whose degree we wish to compute. Divide up the hypercubes on the hyperslice at fixed t into 6 tetrahedra each. Each tetrahedron has four corners at which one value of $\hat{\Theta}_t$ is obtained. Let these values be given by $P_i = (a_0^i, a_1^i, a_2^i, a_3^i)$ where a_0, a_1, a_2, a_3 are the projections of the SU(2) element P_i on the Pauli matrices. The mappings P_i, i = 1, 4, define a tetrahedron on S^3. Pick an arbitrary point P on S^3 (say the point with coordinates (0001) in R^4). Decide whether P is inside or outside the tetrahedron formed by the P's. To do this, first define an ordering of the P_i's which makes the determinant D given below positive.

$$D = \begin{vmatrix} a_0^1 & a_1^1 & a_2^1 & a_3^1 \\ a_0^2 & a_1^2 & a_2^2 & a_3^2 \\ a_0^3 & a_1^3 & a_2^3 & a_3^3 \\ a_0^4 & a_1^4 & a_2^4 & a_3^4 \end{vmatrix} \tag{8}$$

Now compute the four determinants obtained by replacing successively each row of D by the coordinates of P. If and only if all four of the determinants have the same sign, then that sign is added to a counter which is adding up the degree of the mapping Θ_t. It is clear that this procedure will count the number of times the mapping Θ_t wraps around S^3. It is also clear that this procedure only has a chance to work when the volume of the tetrahedron on S^3 can be unambiguously defined. This is possible only if the four points are close together and a little thought reveals that this will happen when the gauge coupling is small (or β is large).

Finally, Q is given by

$$Q = \Sigma_t \deg\Theta_t . \tag{9}$$

Woit used the method just described to measure $\langle Q^2\rangle/N_s$ (where N_s = Number of lattice sites) for the SU(2) theory and showed that this quantity has the correct weak coupling scaling behavior one expects for it. Further, its numerical value was found to be in agreement with expectations from the large N theory. Other definitions[6] for measuring Q in various theories exist in the literature but for SU(2) gauge theory we find the procedure just described[5] most suitable.

II. Simulations with $\Theta \neq 0$

Given that one can measure the topological charge Q(U) of a gauge configuration, there is still a problem in simulating a theory with a non-zero Θ parameter. This stems from the fact that when analytically continuing from Minkowski to Euclidean space, $F\tilde{F}$ picks up an i but F^2 does not. In the Euclidean formulation, the partition function for $\Theta \neq 0$ is given by

$$\tilde{Z}_\beta(\Theta) = \sum_{\{U\}} e^{S_g(U) + i\Theta Q} \tag{10}$$

where S_g is the pure gauge part of the action[7]:

$$S_g = \frac{\beta}{2} \sum_p \mathrm{tr}\, U_p . \tag{11}$$

U_p is the product of links around unit plaquettes, the gauge group is SU(2) and we impose periodic boundary conditions.

The problem alluded to earlier is that the integrand in (10) is complex. It cannot be interpreted as a probability measure. This means that direct Monte-Carlo methods[8] cannot be applied and we are forced to incorporate the effects of Θ on the dynamics in an indirect fashion.

To do the Monte-Carlo simulation, we update the links using only the gauge part S_g of the total action. The factor $e^{i\Theta Q}$ is used as a weight when computing averages over these configurations. In other words, if $\hat{O}_i$ and Q_i are respectively the expectation value of an order parameter $\hat{O}$ in configuration i and the topological charge of that configuration, the expectation value of $\hat{O}$ for $\Theta \neq 0$ is given by

$$\langle \hat{O} \rangle = \frac{\Sigma_i \hat{O}_i e^{i\Theta Q_i}}{\Sigma_i e^{i\Theta Q_i}} . \tag{12}$$

Equivalently, for every Q, we compute a normalized probability function P(Q) which gives the probability of finding a configuration with topological charge Q in the ensemble of configurations at $\Theta = 0$. Then,

$$\langle \hat{O} \rangle = \Sigma_Q P(Q)\ \hat{O}(Q) e^{i\Theta Q} \tag{13}$$

with O(Q) the average of $\hat{O}$ in the sector with charge Q. This procedure is exact in principle. In practice, one expects that it will need a great deal of statistics to give accurate measurements at large values of Θ. From (13) we define the partition function $Z_\beta(\Theta)$ and the free energy density $F_\beta(\Theta)$

$$Z_\beta(\Theta) = \Sigma_Q P(Q) e^{i\Theta Q} = e^{-N_s F_\beta(\Theta)}$$

$$= \tilde{Z}_\beta(\Theta)/Z_\beta(0) . \tag{14}$$

In our simulation, we directly measure $Z_\beta(\Theta)$ by measuring P(Q). We find that Z is hard to measure in volumes bigger than 5^4. Indeed for a 5^4 lattice, we had to do 150,000 measurements of Q to measure P(Q) with acceptable accuracy. We chose to work at $\beta = 2.1$. The reason for choosing such a low value of β which is just at the cross-over point from strong to weak coupling[9] was twofold. First, the lattice sites we can measure on are limited as already mentioned. So we wish to work at small β to keep the correlation length small and measurable. Further, we expect the correlation length to increase with Θ[2]. Thus, given limitations of computer time, one should work at the smallest value of β at which continuum like behavior is observed. The value $\beta = 2.1$ seemed a reasonable compromise.

In Table I we give our data for n(Q), E(Q) for 3^4, 4^4 and 5^4 lattices. P(Q) is obtained from n(Q) by

$$P(Q) = \frac{n(Q)}{\sum_Q n(Q)} \qquad (15)$$

and E(Q) is the average value of $\frac{1}{2}\,\mathrm{tr}U_p$ in the sector with charge Q.

Table I: Data at ß = 2.1.

	3^4		4^4		5^4	
	n(Q)	E(Q)	n(Q)	E(Q)	n(Q)	E(Q)
-10	0	-	0	-	2	0.46520
-9	0	-	0	-	6	0.46862
-8	0	-	1	0.49592	48	0.47051
-7	0	-	2	0.47341	138	0.47138
-6	0	-	9	0.47458	462	0.46922
-5	4	0.47922	55	0.47429	1491	0.46861
-4	14	0.46082	293	0.47181	4024	0.46754
-3	119	0.45816	1195	0.47002	9012	0.46657
-2	597	0.45820	4048	0.46760	17585	0.46545
-1	2890	0.45492	10219	0.46539	26523	0.46491
0	17653	0.44546	18204	0.46238	31268	0.46471
1	2972	0.45594	10453	0.46535	26841	0.46492
2	616	0.45782	4021	0.46799	17300	0.46562
3	114	0.46147	1144	0.47011	9164	0.46642
4	18	0.45696	292	0.47216	3920	0.46720
5	3	0.46815	50	0.47599	1511	0.46850
6	0	-	11	0.47571	516	0.46979
7	0	-	2	0.47533	136	0.47005
8	0	-	1	0.47424	44	0.47043
9	0	-	0	-	8	0.47063
10	0	-	0	-	1	0.47157

In Figure 1, we plot $F_\beta(\theta)$ vs θ for our various lattice volumes. The 'error-bars' pertain only to the 5^4 lattice data. They are computed as follows:

First, the data set was divided into 10 parts. For each Q, the standard deviation σ_Q on n(Q) was computed. The error in n(Q) in the whole sample is then given by

$$(\Delta n(Q))^2 = \sigma_Q^2/10. \tag{16}$$

Finally

$$(\Delta Z)^2 = \frac{\sum_Q (\Delta n(Q))^2 \, e^{2i\theta Q}}{\left(\sum_Q n(Q)\right)^2} . \tag{17}$$

Strictly speaking, one should not call these 'error-bars'. More correctly, they are estimates of the reliability of the extrapolation from $\theta = 0$ to non-zero θ. Near $\theta = \pi$, there are very delicate cancellations and the extrapolation is not very reliable.

III. Results

The following results can be read off from Figure 1:

1) The correlation length increases with θ. It is between 1 and 1½ lattice spacings from $\theta = 0$ to $\theta = 0.2\pi$ and between 1½ and 2 lattice spacings from there on to $\theta \sim 0.8$.

2) There is no phase transition up to $\theta \sim 0.8$. Increased statistics may show a cusp at $\theta = \pi$ indicating a first order transition.

However, a devil's staircase structure or multiple transitions are ruled out.

3) $F_\beta(\Theta)$ looks very much like a pure cosine plus constant. Fits to the form

$$F_\beta(\Theta) = a\,(\cos\Theta + b\cos 2\Theta) + d \tag{18}$$

give

$$a = -5.7 \times 10^{-3},\ b = .038,\ d = -a(1+b). \tag{19}$$

The small value of b indicates that higher harmonics are suppressed and the dilute gas prediction of a pure cosine[2)] is a good approximation.

4) From Witten's arguments in the large N gauge theory[1)], one can relate derivations of $F_\beta(\Theta)$ at $\Theta = 0$ to the η' effective potential. In particular,

$$m^2_{\eta'} = \frac{2N_f}{a^4 f_\pi^2}\left.\frac{d^2F_\beta(\Theta)}{d\Theta^2}\right|_{\Theta=0} \tag{20}$$

and

$$V_4 = \text{coefficient of } \frac{1}{4!}\eta'^4 \text{ in the effective potential}$$

$$= \frac{4N_f^2}{a^4 f_\pi^4}\left.\frac{d^4F_\beta(\Theta)}{d\Theta^4}\right|_{\Theta=0}. \tag{21}$$

Here N_f is the number of light flavors (3 in this case) and "a" is the lattice spacing which can be traded for $\Lambda_{lattice}$ using measurements of the string tension or of $\langle Q^2\rangle/N_s$. Woit has verified (20)[5)]. From (21) we get

$$V_4 = 700 \pm 25 \tag{22}$$

It is dubious whether V_4 will ever be measured experimentally or be important phenomenologically. Nevertheless, it is a non-trivial number of SU(N) gauge theory at large N.

5) It is commonly believed that Θ being a vacuum angle, is not renormalized. There is a different theory in the continuum limit for every value of Θ. Each derivative of $F_\beta(\Theta)$ with respect to Θ is an observable and should scale correctly as a function of β in the continuum limit. This means that

$$F_\beta(\Theta) = F_\beta(0) + f(\beta)g(\Theta) \tag{23}$$

where $f(\beta)$ is the renormalization group function which, in weak coupling, is given by

$$f(\beta) = \frac{\text{constant}}{a^4} = \frac{\text{constant}}{\Lambda_L^4}\left(\frac{6\pi^2}{11}\beta\right)^{\frac{204}{121}} \exp\left(-\frac{12\pi^2\beta}{11}\right). \tag{24}$$

If (23) is correct, then the quantity

$$r = \frac{6(E_\beta(\Theta) - E_\beta(0))}{F_\beta(\Theta) - F_\beta(0)} \tag{25}$$

is the derivative of $\ln f(\beta)$ with respect to β.
Here (see Table I)

$$E_\beta(\Theta) = \frac{\Sigma_Q E(Q)P(Q)e^{i\Theta Q}}{Z(\Theta)} = \frac{\partial F_\beta(\Theta)}{6\partial\beta}. \tag{26}$$

In particular, r should be independent of Θ. From our data, it is easy to check that indeed r is Θ independent for small Θ (up to $\Theta \sim 0.3\pi$) and equals

$$\frac{\partial}{\partial\beta}\ln f(\beta)\Big|_{\beta = 2.1}.$$

The reason we cannot see its behavior for larger Θ is a matter of poor statistics.

IV. Summary and Conclusions

A method to include the effects of the Θ parameter in Monte-Carlo simulations has been described here. The main conclusions are that the dilute gas picture seems to work better than one would expect and that there is no phase transition between $\Theta = 0$ and $\Theta \sim 0.8\pi$ at $\beta = 2.1$ in the SU(2) theory. The method we suggest needs very high statistics to do simulations on large lattices or large β but in principle, works for lattices of all sizes and for large enough β. Perhaps with a variation of other Monte-Carlo methods[10], a direct simulation of the Θ parameter might be possible.

Acknowledgements

I thank the organizers of this workshop for a travel grant that made it possible for me to attend.

References

1) G. 't Hooft, Phys. Rev. Lett. 37 (1976) 8; Phys. Rev. D14 (1976) 3432; Nucl. Phys. B190 (FS3) (1981) 455;
S. Weinberg, Phys. Rev. D12 (1975) 3583;
E. Witten, Nucl. Phys. B149 (1979) 285; ibid B156 (1979) 269.

2) C. Callan, R. Dashen and D. Gross, Phys. Lett. 63B (1976) 334; Phys. Rev. D17 (1978) 2717.

3) M. Peskin, Cornell Univ. Preprint, CLNS 395 (1978), Ph.D. Thesis.

4) P. DiVecchia, K. Fabricius, G.C. Rossi, G. Veneziano, Nucl. Phys. B192 (1981) 392.

5) P. Woit, 'Topological Charge in Lattice Gauge Theory', Princeton University Preprint, May 1983.

6) B. Berg and M. Lüscher, Nucl. Phys. B190 (FS) (1981) 412;
J. Polonyi, Univ. of Illinois Preprint, ILL-TH-83-3, Jan 1983;
M. Lüscher, Comm. Math. Phys. 85 (1982) 39;
Y. Iwasaki and T. Yoshié, Univ. of Tsukuba Preprint, UTHEP-109, 1983.

7) K. Wilson, Phys. Rev. D10 (1974) 2445.

8) N. Metropolis, A. Rosenbluth, M. Rosenbluth, A. Teller and E. Teller, J. Chem. Phys. 21 51953) 1087;
K. Binder, 'Monte Carlo Methods', eds C. Domb and M.S. Green (Academic Press, New York 1976).

9) M. Creutz, Phys. Rev. Lett. 43 (1979) 553; Phys. Rev. D21 (1980) 2308; Phys. Rev. Lett. 45 (1980) 313;
G. Bhanot and C. Rebbi, Nucl. Phys. B180 (FS2) (1981) 469.

10) D. Callaway and A. Rahman, Phys. Rev. Lett. 49 (1982) 613;
M. Creutz, 'Microcanonical Monte-Carlo Simulation, BNL Preprint, 1983;
U. Heller and N. Seiberg, ZAS Preprint, Princeton, 1983;
A. Strominger, 'Microcanonical Field Theory', IAS Preprint, Princeton, 1982.

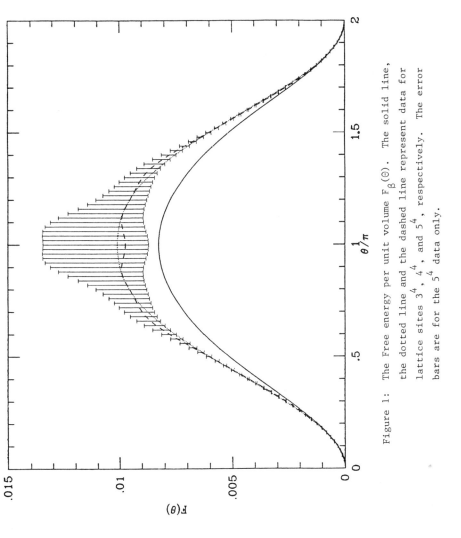

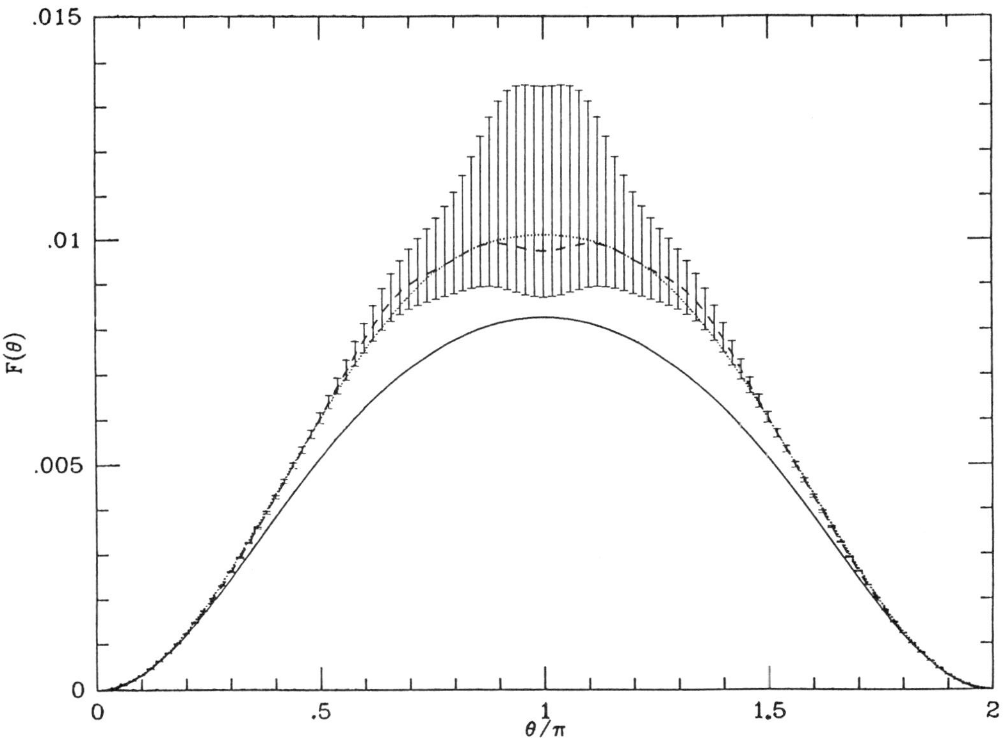

Figure 1: The Free energy per unit volume $F_\beta(\Theta)$. The solid line, the dotted line and the dashed line represent data for lattice sites 3^4, 4^4, and 5^4, respectively. The error bars are for the 5^4 data only.

LOCAL ORDER, GLOBAL DISORDER AND ASYMPTOTIC FREEDOM IN SPIN AND LATTICE GAUGE SYSTEMS

R. Brout

Faculté des Sciences
Université Libre de Bruxelles

Wilson's[1)] program of SU_n pure lattice gauge theory ($n \geq 2$ and no quarks) is comprised of two principal guidelines:

1) There is no deconfining phase transition as the coupling constant (temperature) diminishes.

2) The renormalization group (RG) carries one from weak to strong coupling as the length scale increases (asymptotic freedom). At some minimal length η and beyond this gives rise to the phenomenon of confinement. In quantitative terms, the Wilson loop $< \pi U_i >$ changes over from a perimeter to an area law at the scale η.

In statistical mechanical language these features are couched in terms of order and disorder. On the short length scale there is local order in the sense that if one of the U_i's in the loop is polarized in a certain direction of group space, then there is a gauge in which all the other links are polarized in the same direction, and the value of these $< U_i >$'s is insensitive to the position i on the loop. Then $< \pi U_i > \simeq < U_i >^P$, whence a perimeter law. (One must expect logarithmic corrections due to inevitable small dependence of $< U_i >$ on the position i, even for very small loops). For large loops this is not possible; a link of distance greater than η than the fixed one will have its polarization depending strongly on this distance, in fact exponentially. As an example, consider the case of 2 dimensions, coordinates x and t and take an axial gauge with $U_i = 1$ on links in the x direction. Then for a rectangular loop dimensions R and T, one gets

$$< \prod_{i \in loop} U_i > = [g(R)]^T = e^{-R/\eta T} = e^{-A/\eta}$$

the area law. Presumably in 4 dimensions, the effective coupling constant on the scale of η is sufficiently large that most configurations of plaquettes within loops do not oscillate wildly out of the plane of a coplanar loop, and the above example is relevant. The problem is to show this. (We mention parenthetically that the use of the concept of order in lattice gauge theory has now been legitimized - see refs. 2 and 3)

A good laboratory for this study is the O(n) spin system in 2 dimensions ($n \geq 3$), in that whilst vastly more simple, it does contain many features in common with the gauge problem in 4 dimensions:

(i) For $n \geq 3$, it is commonly admitted, but not yet proven, that there is no phase transition.

(ii) At short distances there is extended local order at low temperatures. This gives rise through the RG to asymptotic freedom[4,5]. On these scales the correlation function, $g(r)$, drops off very slowly-logarithmically.

(iii) At long distances, $g(r)$ drops off exponentially[6].

I shall present here the considerations of W. Deans, A. Silovy and myself[7,8] on these systems emphasizing the following points:

(i) Absence of a phase transition
(ii) Thermal properties
(iii) Asymptotic freedom
(iv) Long distance behavior at low temperature
(v) Transition from short to long distance behavior.
(vi) Possible lessons to be drawn for the gauge problem.

The discussion will be primarily conceptual in character, the aim being a quest for concepts applicable to the gauge problem.

Absence of a phase transition.

Molecular field theory (MFT) always gives a phase transition. We shall show how Onsager's[9,10] reaction field correction eliminates it in 2 dimensions. First here is how MFT works at high temperatures.

The spin Hamiltonian is

$$H = -\frac{1}{2}\sum V_{ij}\vec{S}_i\vec{S}_j - \sum \vec{H}_i^{ext}\vec{S}_i \qquad (\vec{S}_i^2 = 1) \tag{1}$$

MFT is to replace $\vec{H}_{mol}$ by $<\vec{H}_{mol}>$ and to regard it as an external field. Hence correlations are neglected. At high temperatures the response function to a field is $<\vec{S}_i> = (nT)^{-1}\vec{H}_i^{ext}$, so that our approximation gives an integral equation

$$<\vec{S}_i> = (nT)^{-1}<H_{mol}> = (nT)^{-1}\left(\sum V_{ij}<\vec{S}_j> + \vec{H}_i^{ext}\right) \tag{2}$$

solvable by Fourier transforms

$$<\vec{S}_q> = \left(\vec{H}_q^{ext}/nT\right)\left(1 - V(q)/nT\right)^{-1} \tag{3}$$

where $S_q = \sum_i \vec{S}_i \exp qR_i, \quad V(q) = \sum V_{ij} \exp q\cdot(R_i - R_j)$.

By the fluctuation theorem we then get

$$< |S_q|^2 > = \left(1 - V(q)/nT\right)^{-1}. \tag{4}$$

Note the inconsistency; the approximation assumed the spins statistically independent, yet the output yields a correlation function whose Fourier transform is (4). The result is a divergence at $T = T_c = \frac{v(0)}{n}$, a typical second order transition.

The fundamental remark of Onsager is $<\vec{H}_{mol}(i)>$ cannot be the true orienting field spin "i", for as "i" swivels around in group space, its neighbors, which are correlated with it, have one component of polarization with swivels with it. This gives rise to a piece of $\vec{H}_{mol}(i)$ proportional to $\vec{S}_i$ (the reaction field). This piece clearly is ineffective in orienting $\vec{S}_i$ being parallel to it. Onsager's approximation is to replace the MFT molecular field by

$$\vec{H}_{orient}(i) = \langle \vec{H}_{mol}(i) \rangle - \mu \vec{S}_i \quad (5)$$

with $\langle \vec{H}_{mol}(i) \rangle$ given by eq. (1). The single particle energy of i is then $-\vec{S}_i \cdot \vec{H}_{orient}(i)$. At $H_{ext} = 0$, this is equal to 2E/N. Thus we can calculate μ by multiplying (5) by $\vec{S}_i$ and setting $H_{ext} = 0$, in which case $\langle \vec{S}_i \rangle = \langle H_{mol}(i) \rangle = 0$. Thus

$$\mu = -2E/N \quad (6)$$

The rest follows the pattern of MFT

$$\vec{S}_q = \frac{\vec{H}_q^{ext}}{1-(V-\mu)/nT}, \quad \langle |\vec{S}_q|^2 \rangle = \frac{1}{1-(V-\mu)/nT} \quad (7)$$

which combined with (6) gives a self consistent equation for μ

$$\mu = \sum V(q) \langle |\vec{S}_q|^2 \rangle = \sum V(q) \left(1 - \frac{V(q)-\mu}{nT}\right)^{-1} . \quad (8)$$

Note that the inconsistency of MFT in handling correlations has been removed.

Eq. (8) is the spherical model of Berlin and Kac[11)] and is known to be a rigorous solution of the problem at $n = \infty$.[12)] Our presentation has been designed to bring out its physical content. It is in fact a physically appealing approximation and we would be hardly surprised if it turned out to be a viable approximation in the high temperature regime.

The main point is that Eq. (8) eliminates the MFT phase transition at d = 2. The would be critical temperature "T_c" is at "T_c" = $\{v(0) - \mu(T_c)\}/n$ where $\mu(T_c) = n$ "T_c" $\Sigma\{v(q)/v(0)-v(1)\} \sim \int d^2q/q^2$. The infra-red divergence at d = 2, prevents the realization of the transition. Rather a change in behavior occurs below some value of T. At low T, it is convenient to define a mass

$$m^2 = nT - V(0) + \mu \tag{9}$$

so that (8) becomes

$$1 = nT \int \frac{d^2q}{(2\pi)^2} (q^2 + m^2)^{-1} \tag{10}$$

and $m^2 = \exp-(2\pi/nT)$ in inverse (lattice spacings)^2. Thus there is some $T(\equiv T_o)$ above which one gets behavior typical of the first few terms of the high energy expansion and below which one finds exponentially long correlations. In this region we have

$$\mu = -2E = V(0) - nT + O\left(\exp -\frac{2\pi}{nT}\right) \tag{11}$$

giving rise to a specific heat C equal to n/2 at low temperatures.

Herein lies the Schilles heel of the approximation when applied to systems of finite n. For as $T \to 0$, spin wave theory becomes exact and $C \to (n-1)/2$. Therefore it behooves one to analyze the problem anew from the low temperature point of view and to seek out those features which support the conclusion that there is no phase transition. After all the Ising model (n=1) has a transition as well as the case n=2. Is there a value of n which is qualitatively equivalent to n=∞? The answer is yes - due to asymptotic freedom.

True low temperature behavior.

The correlation length at low temperature is much greater than the lattice distance, so that $\vec{S}_i$ varies slowly. This permits a continuum formulation: Take $V_{ij} \neq 0$ for near neighbors only. Then

$$-\sum V_{ij}\vec{S}_i\cdot\vec{S}_j = \sum(\vec{S}_i-\vec{S}_j)^2-2 \to \int(\partial_\mu\vec{S})^2 d^2x + \text{const} \tag{12}$$

$$Z=\int\prod_x d\vec{S}(x)\,\delta(\vec{S}^2(x)-1)\exp -\frac{1}{T}\int d^2x\,(\partial_\mu\vec{S})^2 . \tag{13}$$

Define $\vec{S} = \sigma, \vec{\pi}$, $\vec{\pi}$ an (n-1) dimensional vector, and integrate over σ

$$Z=\int\prod_x \frac{d\vec{\pi}(x)}{\sqrt{1-\pi^2(x)}}\exp -\frac{1}{T}\int d^2x\left((\partial_\mu\vec{\pi})^2+(\partial_\mu\sqrt{1-\pi^2})^2\right). \tag{14}$$

One can then make a systematic expansion in T and as $T \to 0$, the term in $(\partial\mu\sqrt{1-\pi^2})^2$ gives $O(T^2)$ effects only. Thus

$$\lim_{T\to 0} E = \frac{n-1}{2}T\sum_q 1 \tag{15}$$

whence

$$\lim C = \frac{n-1}{2} + O(T).$$

Note however that the low T expansion must be used with caution. For example $<\sigma> \simeq 1 - <\pi^2>/2 = 1 - \frac{n-1}{2}\Sigma(1/q^2)$. This diverges in the infra-red (which is of course the fundamental reason why there can be no ordered phase at d=2). In fact the expansion in T is valid only for $q > \eta^{-1}$. The correction to E and C due to effects $q < \eta^{-1}$ is $O(\eta^{-2})$ hence exponentially small whereas they are essential to calculate $<\sigma>$, to find $<\sigma> = 0$.

We are not prepared to conjecture on the behavior of C as a function of T. At low T, C is equal to (n-1)/2 where (n-1) is the number of active degrees of freedom within a correlation length, η. As T increases the nth degree of freedom starts to activate, even within the correlation length so that C should rise up to its spherical model value, n/2, before falling off in typical high temperature fashion, the latter presumably well approximated by the spherical model. Furthermore, if these ideas are right the good variables to use are C/n vs. nT. At high T the plot should be independent of n. All curves should have a maximum at a value near to 1/2 and at a temperature which in good approximation should be independent of n. All the n dependence has to come on the low temperature side where C/n has a level out at $1/2(1 - \frac{1}{n})$ as $T \to 0$. The integrals under the curves must of course be n independent.

The behavior has been confirmed in preliminary Monte Carlo runs by Dr. J-L. Colot of the University of Brussels who has kindly permitted me to present his data on O(3), O(4) and O(5). The accompanying figure shows the data points. I've not had time to sketch in the spherical model. It would be a curve starting at C = 1/2 at T = 0 and turning over at nT ≃ 2, to follow the other curves. It is noteworthy that all the high temperature data, correctly scaled, are in very good approximation n independent. Elaborate high termperature expansions would seem superfluous as the spherical model seems to account well enough for the facts. We may even note that had we drawn the corresponding curve for the Ising model, there would be a critical point at $nT_c = 2.24$, near the common maximum. Likewise, I expect the U(1) system will fit in as well. All of this will be analyzed and published by Dr. Colot in due course.

We also mention that there is Monte Carlo evidence[13)] on the O(3) system which shows that the third degree of freedom begins to activate at about 1/2 the temperature of the maximum C ($\equiv T_{max}$). At about $T = 0.8\ T_{max}$ it is contributing its full content of 1/2 to C.

We conclude that the known thermal data conform quite well to our understanding of the problem.

Asymptotic Freedom

This is the chapter that is the essential complement to the understanding of the absence of a phase transition. We find Polyakov's original work[4)] closest to our aim of acquiring physical concepts. He considers the RG as follows. One is after the effective interaction of 2 spins after integrating out the configurations of intermediate spins. Let these 2 be fixed in group space and designate them by heavy lines in the accompanying diagram. The average configurations of the intermediate spins is drawn in light lines

The 2 heavy lines fix a plane in group space and the intermediate spins have average values which rotate in this plane. Pass to a "body-fixed" system of axes to parametrize these intermediate spins, $\vec{S}(x)$, by introducing unit vectors $\vec{E}_1(x)$, $\vec{E}_2(x)$, $\ldots\vec{E}_n(x)$. $\vec{E}_1(x)$ is along the mean direction of $\vec{S}(x)$, $\vec{E}_2(x)$ lies in the plane and $\vec{E}_3\ldots\vec{E}_n$ out of it. Then

$$\begin{aligned} \vec{S}(x) &= \vec{E}_1(x)\sigma(x) + \textstyle\sum_{i=2}^{n} \vec{E}_i(x)\pi_i(x) \\ (\partial_\mu \vec{S})^2 &= (\partial_\mu\sigma + A_\mu^{12}\pi_2)^2 + (\partial_\mu\pi_2 + A_\mu^{21}\sigma)^2 + \textstyle\sum_{i\geq 3}(\partial_\mu\pi_i)^2, \\ A_\mu^{12} &= -A_\mu^{21} = \vec{E}_1\cdot\partial_\mu\vec{E}_2 . \end{aligned} \tag{16}$$

At low T, $\partial\mu\sigma$ is negligible and $A_\mu^{12} = \partial\mu\theta$, θ being the angle between the 2 fixed spins and $\partial\mu$ the lattice difference on the big lattice defined by the set of fixed spins. Thus, at low T we get

$$(\partial_\mu \vec{S})^2 = (\partial_\mu\Theta)^2\Big[1 - \sum_{i=3}^{n} \pi_i^2\Big] + \sum_{i=2}^{n}(\partial_\mu\pi_i)^2 . \tag{17}$$

Integrating over intermediate spins is most conveniently performed by passing over to momentum space and integrating over $\pi_i(q)$ for $\Lambda' \leq q \leq \Lambda$ where $\Lambda = 2\pi$ x inverse lattice distance. This gives rise to an effective Lagrangian

$$L_{eff}(\Lambda') = (\partial_\mu\Theta)^2\Big(1 - \frac{(n-2)T}{(2\pi)^2}\Big)\int_{\Lambda'}^{\Lambda} \frac{d^2q}{q^2} \tag{18}$$

whence

$$\frac{1}{T_{eff}(\Lambda')} = \frac{1}{T}\left[1 - T\frac{n-2}{2\pi}\ln\frac{\Lambda}{\Lambda'}\right] \tag{19}$$

$(\frac{n-2}{2\pi})T$ is thus the one loop approximation to the β function. For 2 loops as well as the field theoretic derivation the reader is referred to ref. 5.

Wave function renormalization Z_π follows from $\langle \sigma \rangle = 1 - \langle \pi^2 \rangle/2$ i.e., the spin waves on a certain length scale ride over a local magnetization characteristic of that scale. Thus

$$Z_\pi = 1 - \frac{n-1}{2} T \ln\frac{\Lambda}{\Lambda'} \tag{20}$$

where T is on the scale Λ. At scale q, we have from (19) and (20)

$$\frac{d\ln Z_\pi}{d\ln q} = \frac{n-1}{2\pi}T(q) = \frac{n-1}{2\pi}T\left[1 + T\frac{n-2}{2\pi}\ln\frac{q}{\Lambda}\right],$$

$$Z_\pi = \left[1 + \frac{n-2}{2\pi}T\ln q\right]^{(n-1)/(n-2)}$$

and

$$\langle \pi^2(q)\rangle = \frac{Z_\pi(q)T(q)}{q^2} = \frac{T}{q^2}\left[1 + \frac{n-2}{2\pi}T\ln q\right]^{1/(n-2)} \tag{21}$$

All formulae take on validity for $(\frac{n-2}{2\pi})$ T log q < 1. The important point is that for n ≥ 3, the temperature grows with length scale until T(q) = 0(1). Thus there is a low temperature disordering mechanism due to infra-red effects of the same type which eliminated the phase transition from the high temperature reaction field analysis, provided n ≥ 3. In this circumstance we gain confidence, but of course have not proven, that the earlier conclusion of no phase transition is correct provided n ≥ 3. At the same time there is suggested an approach to an estimate of the correlation length at low T.

Long distance behavior at low T.

The renormalization group RG analysis breaks down at $T(q)=0(1)$. From thermal data we see that σ loosens up to give its full contribution to C at $T = T_1$ ($\simeq 0.6$ for 0(3)). Therefore we use the RG result up to length scale ξ where $T(\xi) = T_1$. Since the temperature range $T \geq T_1$ is adequately described by the spherical model, we use the latter to describe the correlation function from ξ on out, and at fixed temperature T_1, i.e. we construct block spins ξ on a side, to yield a near neighbor 0(n) model of temperature T_1 and approximate this effective model by the spherical model. This yeilds the correlation length $\xi(T)$ in two loop approximation to the β function

$$\xi(T) = \exp \frac{2\pi}{nT} \left(\frac{1 + 2\pi/T_1}{1 + 2\pi/T} \right) \exp\left[\frac{2\pi}{n-2} \left(T^{-1} - T_1^{-1} \right) \right] . \tag{22}$$

The first factor is the spherical model value of ξ at T_1 and the second factor is due to the RG analysis. For n=3 and $T_1 = 0, 6$, one finds $(T) = 0.95 \times 10^{-2} \exp\{2\pi/(n-2)T\}$ as compared to Shenker and Tobochnek's Monte-Carlo estimate[6] of the prefactor ($10^{-2} \pm 30\%$). The precision of the estimate we have given should not be taken too seriously until the choice of T_1 is put on a firmer theoretical footing. From the thermal data it is clear that $0.5 \leq T_1 \leq 0.75$, the lower bound being where the nth degree of freedom starts taking off and the upper where the characteristic high temperature descent of the specific heat begins - a sign of total disorder. (Once this happens there will be no further renormalization of temperature, an effect characteristic of extended local order.) Corresponding to the range of the prefactor of ξ will be between 1/2 and 2×10^{-2}. We conclude that we have a good semi-quantitative grasp of what is going on.

The transition region and bound state formation.

We now wish to report on work in progress on the transition region, our attempt to describe in dynamical terms the activation of the nth degree of freedom.

Our source of inspiration is the bound state mechanism of Bardeen, Lee and Schrock (BLS)[14]. These authors examine the problem in $2+\varepsilon$ dimensions where there is a phase transition for $\varepsilon > 0$. Further, they consider the $n = \infty$ limit where the one loop approximation (OLA) becomes rigorous. They use the formulation eq. (14) but change over to a field $\vec{\phi}$ related to $\vec{\pi}$ by a stereographic transformation $\vec{\pi} = \vec{\phi}/(1+\phi^2/Y)$. (Whereas $\pi^2 < 1$, $\vec{\phi}$ is a more convenient field since $-\infty \leq \phi_i \leq -\infty$). In terms of $\vec{\phi}$ one has

$$L = \frac{1}{2T} \frac{(\partial_\mu \phi)^2}{(1+\phi^2/4)^2} \tag{23}$$

with a Jacobian in the measure given by $\pi\ \{1+\phi^2/Y\}^{n-1}$. In one loop approximation one has for the ϕ propagator

$$\Delta_\phi = \frac{T[1+\langle\phi^2\rangle/4]^2}{q^2 - [\langle(\partial_\mu\phi)^2\rangle/2\,(1+\langle\phi^2\rangle/4)]} \tag{24}$$

where

$$\langle\phi^2\rangle = (n-1)\int \Delta_\phi \frac{d^d q}{(2\pi)^d} \tag{25}$$

$$\langle(\partial_\mu\phi)^2\rangle = -m^2\langle\phi^2\rangle\ . \tag{26}$$

Eq. (26) is established either by dimensional or lattice regularization. The latter making use of the Jacobian in the measure of the field integration. From (26) and (24) follows

$$m^2 = m^2 \frac{\langle\phi^2\rangle/2}{1+\phi^2/4} \tag{27}$$

whereupon either $m^2 = 0$ ($T < T_c$) or $m^2 \neq 0$ ($T > T_c$) in which case $< \phi^2 > = 4$ (or $< \pi^2 > = 1$ in OLA). This gives back the spherical model with n replaced by (n-1).

The important discovery of BLS is that the eigenvalue condition on the mass ($< \phi^2 > = 4$) gives rise to a bound state and they prove

$$\langle (\phi^2(q))^2 \rangle - \langle \phi^2(q) \rangle^2 = \frac{64T}{q^2 + m^2} \tag{28}$$

implying in OLA, $< \sigma^2(q) > = T/q^2 + m^2$. The σ is reinstated as a degree of freedom completely equivalent to the π. The 2 particle cut cancels out of the propagator of $\phi^2(x)$. In this way O(n) is restored in the disordered phase.

BLS also study the ϕ ϕ scattering amplitude and in general the 2-particle cut is present, it only cancels out when the initial and final pairs of ϕ's are at a common point.

We are now attempting to use these ideas at d = 2 and n finite. Consider the following heuristic argument. Take a finite sample of length L on a side and suppose the system is disordered ($m^2 \neq 0$). Then from BLS, it is required that

$$(n-1) \sum_{q > L^{-1}} \langle \pi^2(q) \rangle = 1 \quad . \tag{29}$$

Upon substituting Eq. (21), one finds (up to subtle effects involving the Shenker-Tobochnik prefactor of 10^{-2}) that the condition (29) can be satisfied only for $L \geq \eta$ i.e. small boxes are magnetized and the eigenvalue condition giving rise to a bound state cannot be met. Only (n-1) degrees of freedom/spin are active. For big boxes, Eq. (29) can be met by introducing a mass into the π propagator. This allows for the bound state, the system is disordered and the spherical model is recovered on length scales greater than η.

(Added note: In the course of the question session following this talk, Professor Lüscher challenged the above argument on the basis of his own work (Physics Lett. 118B, 391 (1982)) in which the low lying mass spectrum of the Hamiltonian is analyzed, Hamiltonian in the sense of the log of the transfer matrix, a one dimensional object). The states in the spectrum are in O(n) multiplets and the first excited state determines the correlation length. This is the O(n) vector multiplet $\vec{\pi}$ and σ. Professor Lüsher therefore states that a finite sample cannot be ordered, thereby breaking O(n) symmetry.

My answer is that strictly speaking, this is correct. No finite system can exhibit spontaneous broken symmetry. It is a matter, however, of weighing the importance of the accessibility "other classical vacuum". One must bear in mind that the sample I am considering is finite in both dimensions, whereas Lüscher studies samples which are finite in one dimension but tending to infinity in the other (thus allowing him to consider only low lying states of the spectrum). To describe a box whose length is less than η in both directions requires the complete spectrum. A magnetized small box is a wave packet that can be built up of all these states. True, this packet will dissipate after a while - but I presume a long while - something like a time scale proportional to the exponential of the number of particles in the box - and thus can be chosen arbitrarily large as T becomes arbitrarily small. Thus the Monte Carlo runs at low temperatures where the sample size is less than η are done on magnetized systems. One may appreciate the problems as follows. Fix the polarization of one spin in the sample. Then all spins will be net polarized in the same direction if $L < \eta$. Surely the properties which are being described cannot depend on fixing this single spin. Further if one releases the fixing mechanism one must wait a time for the message to be relayed to a finite fraction of all spins in the box for the magnetization to change its orientation. This will be a very slow process so that all but the excitation modes of wave-length of O(L) will simply ride on the total magnetization as it moves about. These will be the (n-1) degrees of freedom/spin we are talking about.

In fact I consider it gratifying that Professor Lüscher has successfully accounted for the correlation length in terms of the O(n) vector multiplet. This is precisely the point I am making - on the long distance scales σ is restored, due to bound state formation of the π's.)

Our heuristic argument suggests that the radius of the bound state is η (since it does not exist in small boxes, but does exist in large ones). How to get a quantitative line on this is the subject of our current research. The idea is that on the large length scale, the renormalized π's of the RG analyses are really blocks of unrenormalized π's, hence they are non-local objects constructed from the original fields. If the nonlocality is less than η , then the relative momentum transfer of $\pi\ \pi$ scattering is large compared to the mass and the 2-particle cut should be an important part of the scattering. Whereas on the scale greater than η , the momentum transfer is small and the scattering will proceed through the one-particle (σ) pole. This touches on the true difficulty of the problem. The states $\vec{\pi}$, σ of the spectrum are non locally related to the fields $\vec{\pi}$.

<u>Lessons drawn for lattice gauge theory (LGT)</u>.

(i) Firstly why is there no phase transition from the reaction field point of view. In LGT one encounters a first order phase transition in molecular field theory. This is accomplished by a limit point of metastability at a temperature T_M where all masses vanish, but the magnetization is finite. Once more infra-red divergences prevent this occurrence. Therefore in extrapolating from high to low temperature, one cannot encounter this limiting situation; the ordered phase is not attainable.

(ii) The specific heat is known to peak and the explanation seems similar to that given here. In fact, the temperature of the peaking seems rather close to T_M.

(iii) We expect the glueball to a sort of (n^2)th state, a SU_n singlet bound state of gluons analogous to the BLS σ. Its radius should be the confinement length scale. Of course one cannot expect degeneracy with the gluons because of confinement and screening of SU_n nonsinglets.

(iv) Finally there may be a general simple approximation in which quantities scale like n (or n^2) to describe high temperature or large length scale phenomena. This reminds one of Professor Mack's talk at the Bonn conference three years ago in which he suggested that the area law of SU_n theories could be built up from Z_n.

References

1. K. Wilson, Phys. Rev. D10 (1974) 2445.
2. E. Brézin and J.M. Drouffe, Nucl. Phys. B200 {FS4}(1982) 93.
3. D. Brout, R. Brout and M. Poulain, Nucl. Phys. B205 {FS5}(1982) 23.
4. A.M. Polyakov, Phys. Lett. 59B (1975) 79.
5. E. Brézin and J. Zinn-Justin, Phys. Rev. B14 (1976) 3110
6. S. Shenker and J. Tobochnik, Phys. Rev. B22 (1980) 4462.
7. R. Brout and W. Deans, Nucl. Phys. B215 {FS7} (1983) 407.
8. R. Brout, W. Deans and A. Silovy, Phys. Rev. (in press).
9. L. Onsager, J. Am Chem. Soc. 58 (1936) 1486.
10. R. Brout and H. Thomas, Physics 3 (1967) 317.
11. T. Berlin and M. Kac, Phys. Rev. 86 (1952) 821.
12. H.E. Stanley, Phys. Rev. 176 (1968) 718.
13. R.E. Watson, M. Blume and G.H. Vineyard, Phys. Rev. B2 (1970) 684.
14. W. Bardeen, B. Lee and R. Schrock, Phys. Rev. D14 (1976) 985.

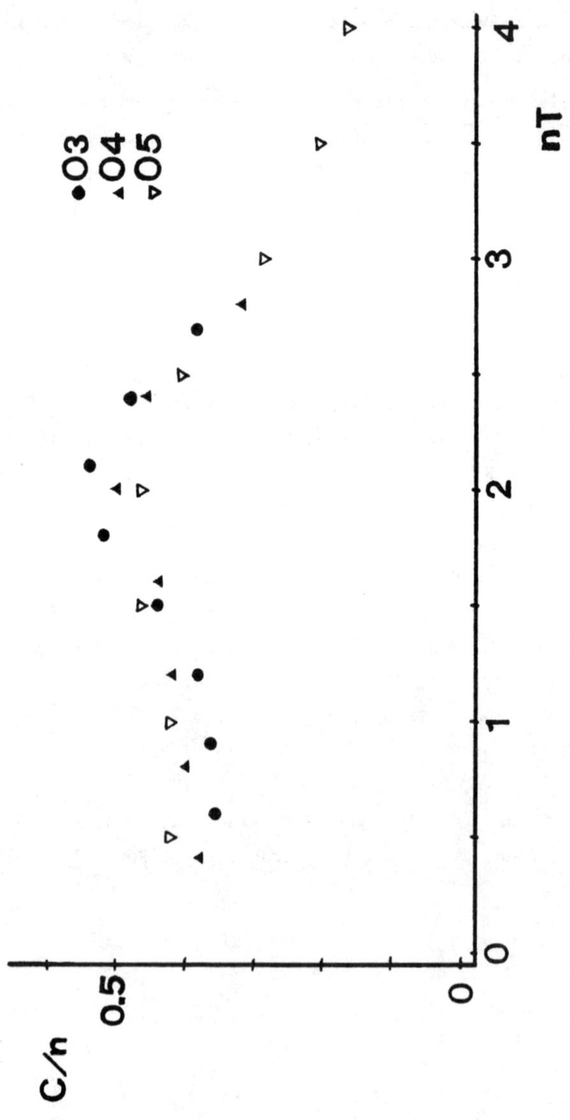
C/n
0.5
0
nT
0
1
2
3
4
.03
.04
.05

ASTROPHYSICAL IMPLICATIONS OF THE RUBAKOV EFFECT

W. Nahm

Physikalisches Institut der Universität Bonn
Nussallee 12
D-5300 Bonn 1

1. Introduction

Together with the baryon density of the universe and the spontaneous decay of nucleons, the properties of GUT monopoles might be among the few clues to physical phenomena at the energy scale of 10^{15} GeV, where all known gauge interactions become of comparable strength. Thus the direct or indirect detection of magnetic monopoles is one of the main challenges for physicists and astrophysicists.

Monopoles are rare and theoretical considerations don't give very useful clues for their absolute number in the universe. However, it should be possible to calculate their relative abundance at various places in our galaxy. Because of their low density, interactions between monopoles are usually not important, such that the relative abundances are independent of the total number. The one exception concerns the interiour of neutron stars, where the annihilation of oppositely charged monopoles may be common.

A priori, monopoles may be seen individually by laboratory effects or indirectly by cumulative effects of large numbers of them. A typical cumulative effect is the exhaustion of magnetic fields from which the monopoles take up energy. Much more hypothetical is the creation of galactic magnetic fields by density fluctuations in a monopole plasma[1)], or a contribution of monopole annihilation products to the spectrum of energetic cosmic rays[2)].

A very sensitive way to detect monopoles both in the lab and by cumulative effects should be the observation of the Rubakov effect, i.e. the monopole catalysis of nucleon decay with a typical strong interaction cross section[3)]. Due to this effect passing monopoles may be seen by experiments designed to detect spontaneous nucleon decays. Moreover, the effect may be visible or even conspicuous in astronomical objects which had enough time and stopping power to collect large numbers of monopoles from the galactic flux, in particular old neutron stars, the sun, and perhaps Jupiter.

2. The Rubakov Effect

As it has been claimed that the effect only takes place with a very small weak cross section or not at all, it may be useful to give an elementary description.

The magnetic charge of a monopole is given by Dirac's quantization condition

$$eg = 2\pi . \tag{1}$$

Multiply charged monopoles are expected to be unstable under decay into monopoles of charge +g or -g.

A quark in the neighbourhood of a purely electromagnetic monopole would violate the quantization condition, such that the monopole also must have a colour-magnetic charge q_c and a corresponding colour-magnetic field extending out to a screening distance of the order of 1fm. We may use this field to define the 3-direction in colour space. For fermions with electric charge q_e and colour hypercharge q_c one must have

$$q_e g + q_c g_c = 2\pi N \tag{2}$$

with integer N. To yield a stable monopole, g_c must be minimal. The unique solution of this equation yields N=1 for e^+, $\bar{d}_3$, u_1, u_2 and N=0 for d_1, d_2, u_3, and of course the neutrino. For the antiparticles, N changes sign. The same pattern repeats for the higher generations.

GUT monopoles are given by an embedding of a subgroup $SU(2)_M$ into the grand unified group. Under this subgroup the fermions with N=0 are singlets, the others assemble into doublets. Within each doublet the charge difference must be the same. The doublets might mix the generations, but if this is not the case, one has

$$\begin{pmatrix} e^+ \\ d_3 \end{pmatrix}_L \quad \begin{pmatrix} \bar{d}_3 \\ e^- \end{pmatrix}_L \quad \begin{pmatrix} u_1 \\ \bar{u}_2 \end{pmatrix}_L \quad \begin{pmatrix} u_2 \\ \bar{u}_1 \end{pmatrix}_L \tag{3}$$

and the corresponding right-handed anti-particles. In particular, this result applies to the SU(5) GUT.
The Rubakov effect is due to the Adler-Bell-Jackiw anomaly. For each left-handed fermion field one has

$$\frac{dn}{dt} = \frac{1}{8\pi^2}\int (qE)(qB)\, d^3x + A_{int} . \tag{4}$$

Here n is the fermion number, and the gauge field term on the r.h.s. includes electromagnetism and colour hypercharge, i.e.

$$qB = q_e B_e + q_c B_c \tag{5}$$

and similarly for qE. The term A_{int} comes from interaction terms in the Lagrangian describing processes which change the fermion under consideration into a different one. Among them we have the mass term

$$A_m = \int \bar{\psi} \varphi \psi \, d^3x \tag{6}$$

where φ is a field whose vacuum expectation value is the fermion mass.

Under most circumstances, the r.h.s. of eq. (4) is dominated by

A_m, and the equation only indicates that in the presence of a mass term the helicity is not well defined. Ordinary electric and magnetic fields are too weak to see an effect. However, a charge q' in the presence of a magnetic monopole induces a conspicuous effect. If the distance of the charge from the monopole is r, one finds

$$\frac{1}{8\pi^2}\int (qE)(qB)\, d^3x = (qq')N r^{-1} \tag{7}$$

with N given by eq. (2). If

$$r \ll (qq')\, m^{-1} \tag{8}$$

the mass term is negligible. For such distances the low energy physics responsible for the pairing of e_L^-, e_L^+ etc. is irrelevant, and chirality gets the same status as flavour.

The remaining interaction terms are small, too, with the exception of the $SU(3)_c$ term and a contribution from the fermion interaction with the off-diagonal component of the $SU(2)_M$ gauge field. This field is condensed in the core of the monopole. As the core radius is large compared to the monopole Compton wavelength, a classical description is possible.

The off-diagonal gauge field components change the fermions of each doublet into each other. If one adds for a given doublet the two eqs. of type (4), the corresponding interaction terms cancel. Adding up all colour components cancels the remaining colour interaction terms, such that one finally obtains

$$\frac{d}{dt}\left(n(e_L^+) + n(d_L)\right) \approx (\bar{q}q')\, r^{-1} \tag{9a}$$

$$\frac{d}{dt}\left(n(e_R^+) + n(d_R)\right) \approx -(\bar{q}q')\, r^{-1} \tag{9b}$$

$$\frac{d}{dt}\left(n(u_L) - n(u_R)\right) \approx 2(\bar{q}q')\, r^{-1} \tag{9c}$$

and analogously for the higher generation. Here

$$\bar{q} = q(e^+) - q(d_3) - 2q(u_1) = \left(\tfrac{4}{3}e, -\tfrac{2}{3}e_c\right). \tag{10}$$

The reactions changing the fermion type can be described by an interaction of the fermion of charge q' with a condensate. At distance r the condensate contains a fermion from each $SU(2)_M$ doublet for which condition (8) is satisfied. The condensate does not contain charged condensed gauge or Higgs bosons outside its core, such that one has an ordinary conservation equation for the electromagnetic current. The same remark applies to $SU(3)_c$. If only the fermions of the first generation are involved, the condensate therefore has the components $(e^+ \bar{e}\, \bar{u}\bar{u})_L$, $(\bar{d}\bar{d}\bar{u}\bar{u})_L$, and $(e^- duu)_L$. At the monopole core they mix because of the $SU(2)_M$ gauge field condensate. In general, core physics yields the boundary conditions which fix the relative contributions of the various components of the condensate.

From this description of the Rubakov effect it should be clear that the weak interaction scale is rather irrelevant. Weak interaction effects only appear in small terms which we neglected. In particular one may calculate in the limit $m_W \to \infty$, where e.g. the GUT group SU(5) is reduced to SU(4) without changing the effect. Furthermore, mass effects are irrelevant, once the distance is small enough. If the cross section is written in the form

$$\sigma = \sigma_0 \cdot 0.1\, fm^2 / v \tag{11}$$

we obtain a σ_0 of order 1, rather than of order 10^{-3} as obtained in ref. 4.

The expected nucleon decay modes are mainly

$$\begin{aligned} p &\to e^+\pi^0 \\ &\to e^+\pi^+\pi^- \\ &\to e^+\pi^0\mu^+\mu^- \end{aligned}$$

$$\begin{aligned} n &\to e^+\pi^- \\ &\to e^+\pi^0\pi^- \\ &\to e^+\pi^-\mu^+\mu^- \end{aligned}$$

where the muonic modes come from the condensate $e^+\bar{d}\bar{u}\bar{u}\mu\bar{\mu}$.

3. Monopole Production

A lower bound for the number of GUT monopoles forming in the phase transition which breaks the GUT symmetry is obtained from the causal horizon[5]. Usually the argument is formulated in the following way: At the time of the condensation at scales beyond the causal horizon, the orientation of the Higgs field was random. This yields roughly one topological defect in the condensed Higgs field per causally connected volume. The argument is not quite conclusive, because the Higgs orientation is a gauge dependent quantity, but one may formulate it in some hyperaxial gauge. Alternatively, one may look at elements of the holonomy group given by the gauge field: Let x_o be a fixed base point and consider standard, e.g. rectangular loops from x_o through a variable point x and back. This yields a function g(x) which takes values in the gauge group. Before the GUT phase transition this function is random at sufficiently large distances, and at distances beyond the horizon scale it will remain random even afterwards. At smaller distances it can relax to take values in the unbroken subgroup of G, but beyond the horizon there are obstructions so such a relaxation which will develop into magnetic monopoles.

For a big bang dominated by radiation the temperature at time t is of order $(m_{pl}/t)^{1/2}$ with $m_{pl}=10^{19}$ GeV, and the distance to the horizon is of order t. If the monopoles form at temperature T the ration of monopole to photon density is

$$\frac{n_M}{n_\gamma} \sim \left(\frac{T}{m_{pl}}\right)^3 . \tag{12}$$

For T of order 10^{15} GeV monopoles should be almost as abundant as baryons, which is absurd. However, there are two ways out. One may treat T as a free parameter. This is physically plausible, as a realistic GUT theory must contain many free parameters describing non-renormalizable interactions which are suppressed by powers of m_{pl} and only will be of order one at energies where quantum gravity effects are important[6]. If T is less than 10^{10} GeV, monopoles abundance is compatible with present observations.

A second idea to explain the rarity of monopoles starts with the question, why we live in a homogeneous part of the world which has an entropy of order 10^{85}, or more if the universe stretches much beyond the present horizon. For random solutions of Einstein's equations one rather obtains 1 than 10^{85}. That large an entropy may have been assembled slowly during a succession of big bang during a first order phase transition[8]. If this happened before monopole formation, it is irrelevant for the monopole density, but if the GUT phase transition was involved, the monopoles may have been diluted by the newly created entropy, and may be impossible to find.

Both scenarios face serious problems, but they indicate at least that the total monopole number of our galaxy should be treated as a free parameter.

4. Monopoles in Neutron Stars

In our galaxy, ordered magnetic fields of about 10^{-6} Gauss stretch across distances of about 300pc. They accelerate monopoles of mass $\mu\cdot10^{16}$ GeV to a velocity of order $3\mu^{-1/2}\cdot10^{-3}$c. This applies to monopoles of mass less than 10^{17} GeV. For higher masses gravitation can keep the monopoles bound to the galaxy, with orbit velocities of 10^{-3} c.

Monopoles arriving at our galaxy may have been ejected from other galaxies with a velocity of the same order, or perhaps belong to a monopole population of interstellar space which did not take part in a gravitational collaps forming a galaxy and now should have a very low temperature. But in our galaxy they would be accelerated by the magnetic fields to the velocities discussed before.

Neutron stars are very efficient collectors for this galactic flux, they absorb every monopole which falls down to their surface[9]. Their collecting area is

$$A \approx \pi R^2 \left(1 + \left(\frac{v_{esc}}{v}\right)^2\right), \tag{13}$$

where $R \sim 10^6$ cm isthe neutron star radius, v is the galactic monopole velocity, and $v_{esc} \sim .4c$ is the escape velocity from the surface of the neutron star. Thus A is of order $10^{18}\mathrm{cm}^2/4\pi$. Consequently the Parker flux limit[10] of $10^{-16}\mathrm{cm}^{-2}\mathrm{sec}^{-1}\mathrm{ster}^{-1}$, which comes from the constraint that the monopoles have been not numerous enough to exhaust the galactic magnetic field, corresponds to a flux $F=10^2\mathrm{sec}^{-1}$ of monopoles falling onto the neutron star.

If there are N monopole in a neutron star of neutron density $.2\mathrm{fm}^{-3}$, the catalysis sets free an energy of order

$$L \sim 10^{19} N \sigma_0 \ \mathrm{erg/sec}. \tag{14}$$

If the neutron star contains condensed pions or quark matter inside, only 10^{-2} of that energy may be emitted as photons, the remainder as neutrinos[11]. Searches show that the visible energy flux of neutron stars must be less than 10^{31}erg/sec[12]. This translates into a limit of $10^{14}/\sigma_0$ for the number of monopoles in their core.

Very strong limits for the galactic monopole flux have been inferred from this limit[9]. However, monopole annihilation was not correctly taken into account. It was assumed that magnetic fields in the neutron star core would prevent significant annihilation[13]. But Kuzmin and Rubakov point out that for a monopole number larger than $\mu^{-3} \cdot 10^7$ the magnetic fields in the core are exhausted by energy transfer to the monopoles[14]. They can not be restored by diffusion from peripheral regions because of the high conductivity of neutron star matter.

For a flux F of monopoles falling to the neutron star, the time to exhaust the field is

$$t_{exh} \sim \mu^{-3/2} \cdot 10^{12} (F/\mathrm{sec})^{-1/2}. \tag{15}$$

Thus for fluxes of less than $\mu^{-3} \cdot 10^{-10}\mathrm{sec}^{-1}$ annihilation will be negligible, but for higher fluxes Kuzmin and Rubakov find that the monopole number reaches an equilibrium value

$$N_{eq} \sim 3\mu^{1/2} \cdot 10^{13} (F \cdot \mathrm{sec})^{1/2}. \tag{16}$$

For neutron stars with superconducting core the physical phenomeny are drastically different, but the final numbers are similar. Monopoles with a mass less that 10^{17}GeV will not be able to penetrate the core. When they collect at the surface of the core their annihilation rate may be high. However, the superconductivity is of second kind, and initially the core will contain about 10^{31} flux lines, where the monopoles can hide. The time needed to destroy these flux lines depends on the nature of the superconducting condensate. Harvey[13] considered a pp condensate, which has flux lines of flux g/2. When a monopole falls into such a flux line, the flux reverses, and the line will annihilate with a neighbouring one. However, this process takes more than 10^2 years, and monopoles are evidently much too rare to exhaust the flux.

In contrast, Kuzmin and Rubakov considered an np condensate with flux lines of flux g. In this case flux lines are annihilated immediately, when a monopole falls into it. Monopoles cross the core in 10^{-4}sec, such that 10^{10} monopoles corresponding to a flux $F=10^{-7}\text{sec}^{-1}$ are sufficient to destroy 10^{31} flux lines in old neutron stars. For higher fluxes, the monopole number reaches a maximum of

$$N_{exh} \sim 3\cdot 10^{13}\,(F\cdot sec)^{1/2} \tag{17}$$

at the time of the flux exhaustion. Then monopole annihilation on the surface of the superconducting core sets in, and the monopole number settles to an equilibrium value

$$N_{eq} \sim 3\mu^{-1}\cdot 10^{12}\,(F\cdot sec)^{1/2}. \tag{18}$$

Kuzmin and Rubakov argue that an np condensate is more likely to form than a pp condensate, as np and pp interactions are of comparable strength, but protons are less numerous. Actually the condensate may contain a pp component, but all flux lines still would have flux g and could be destroyed easily.

5. Monopoles in the Solar System

The collecting area of the Sun for galactic monopoles is about $3\cdot 10^{23}\text{cm}^2/4\pi$, but not all monopoles arriving at the solar surface will be caught[15]. The flux of monopoles ending up in the Sun should be of order

$$F_{Sun} \sim 3\cdot 10^{4}\,F_{neutron\ star}\ . \tag{19}$$

The density of the Sun is $3\cdot 10^{-15}$ times a typical neutron star density, and the energy production per monopole is suppressed by the same factor. Thus even for a flux corresponding to Parker's limit the contribution to the solar luminosity is less than 10^{-5} and unobservable. However, catalysed nucleon decays may be detected by finding neutrinos from pions which decay in flight, in particular from the decay chain.

$$n \to e^+ \pi^- \\ \quad \hookrightarrow \mu^- \bar{\nu}_\mu \, . \tag{20}$$

Only about 1 % of the pions will decay in flight, but still at the Earth one obtains a flux of neutrinos of several hundred MeV with an intensity of order

$$\phi_{\nu_\mu} \sim 30\, \sigma_0 \; cm^{-2} sec^{-1} \cdot \frac{\phi_M^{galactic}}{\phi_M^{Parker}} \, . \tag{21}$$

Conceivably it may just lie above the cosmic ray background.

The energy loss of GUT monopoles in rock is too small to allow a significant monopole collection by the Earth. Nevertheless a limit has been given for the monopole number of the Earth, based on the geothermic flux of $3 \cdot 10^{20}$erg/sec[16]. Such a flux would be caused by $2 \cdot 10^{15}/\sigma_0$ monopoles catalyzing nucleon decay. A somewhat better limit comes from the decay chain (20). Again about 1 % of the pions would decay in flight, such that an energy flux of $3 \cdot 10^{20}$erg/sec would be accompanied by a flux of energetic muon neutrinos of $3 \cdot 10^2 cm^{-2} sec^{-1}$, more than an order of magnitude higher than the cosmic ray backgound.

Jupiter might collect some galactic monopoles and certainly would sweep up solar system monopoles reaching his orbit[15]. At this distance from the Sun orbiting monopoles would have a velocity of order $4 \cdot 10^{-5}$c and an energy of $\mu \cdot 10^{-7}$GeV. Falling down to Jupiter they would reach $v_{esc} = 2 \cdot 10^{-4}$c. To be caught gravitationally they would have to lose about $\mu \cdot 1 MeV cm^{-1}$, which certainly is possible[17]. Jupiter's collecting area for galactic monopoles is $2 \cdot 10^{21} cm^2/4$, for orbiting monopoles $5 \cdot 10^{22} cm^2/4$. A flux

$$F_{Jupiter} \sim \eta \cdot 10^{-3} F_{Sun} \tag{22}$$

where η is perhaps of order 1 translates into an energy liberation by nucleon decay catalysis of about

$$L_{Jupiter} \sim 2 \eta \sigma_0 \cdot 10^{25} \, erg \; sec^{-1} \cdot \frac{\phi_M^{galactic}}{\phi_M^{Parker}} \, , \tag{23}$$

compared to the observed flux of internally generated energy of $3 \cdot 10^{25}$erg/sec.

6. Conclusion

The Rubakov effect should allow detection of magnetic monopoles by presently achievable experimental facilities, if the galactic flux is not much lower than the Parker limit. By a strange conspiracy of physical effects the limits from the luminosity of both ordinary and superconductive neutron stars, from solar muon neutrinos, and from Jupiter's thermal flux are close to the Parker limit. If the thermal flux of Jupiter and Saturn is not due to monopoles, better results for the luminosity of old neutron stars offer the best hope to see or limit monopole effects, as the luminosity only varies with the square root of the flux.

References

1) M.Turner et al., Phys.Rev. D26 (1982) 1296.
   E.Salpeter et al., Phys.Rev.Lett. 49 (1982) 1114.
2) P.Rotelli, Tsukuba preprint KEK-TH 59, Jan. 1983.
3) V.Rubakov, Soviet Phys. JETP Lett. 33 (1981) 644 and Nucl.Phys. B203 (1982) 311;
   C.Callan, Nucl.Phys. B217 (1983) 391.
4) F.Bais et al., CERN preprint TH-3383, Aug. 1982.
5) T.Kibble, J.Phys. A9 (1976) 1387.
6) D.Nanopoulos et al., Phys.Lett. 124B (1983) 171.
7) D.Dicus et al., Sci.Am 248 (1983) 74;
   V.Petrosian, Nature 298 (1982) 805.
8) A.Guth, Phys.Rev. D23 (1981) 347.
9) E.Kolb et al., Phys.Rev.Lett. 49 (1982) 1373;
   S.Dimopoulos et al., Phys.Lett. 119B (1983) 320.
10) E.Parker, Ap.J. 160 (1970) 383.
11) K.Van Riper and D.Q.Lamb, Ap.J. 244 (1981) L13.
12) F.Córdova et al., Ap.J. 245 (1981) 609;
    G.Reichert et al., Ap.J. to be published.
13) J.Harvey, Monopoles in neutron stars, Princeton preprint, 1983.
14) V.A.Kuzmin and V.A.Rubakov, Phys.Lett. 125B (1983).
15) K.Freese and M.Turner, Phys.Lett. 123B (1983) 293.
16) M.Turner, Chicago preprint EFI 82-55 (1982).
17) D.Groom et al., Phys.Rev.Lett. 50 (1983) 573;
    S.Drell et al., Phys.Rev.Lett. 50 (1983) 644.

TOPOLOGICAL OBJECTS, NON-INTEGER CHARGES AND ANOMALIES

Y. Frishman *)

CERN -- Geneva

CONTENTS

---

*) On leave from the Weizmann Institute, Rehovoth, Israel.

TOPOLOGICAL OBJECTS, NON-INTEGER CHARGES AND ANOMALIES

Y. Frishman*)

CERN
CH - 1211 Geneva 23
SWITZERLAND

1. Introduction and Conclusions

Although the talk consisted of a review of the subject, we will concentrate here only on one aspect of the field. This involves the derivation of the total charge in two and four dimensions, thereby also demonstrating why the result is identical for solitons in two dimensions and monopoles in four dimensions.

The method involves a chiral rotation, which brings us naturally to splitting the induced current into two parts, one which is explicitly given and another which is shown not to contribute to the total charge. An explicit calculation of the induced total fermionic current on the topological object is thus avoided. Special care must be given to the method of regularization, in demonstrating that the part of the current we do not compute actually does not contribute to the total charge. We use a different regularization from Goldstone and Wilczek[1]. Our vector current is conserved also when axial vector as well as vector sources are present, while they maintain the O(4) symmetry instead.

2. Non-Integer Charges and Anomalous Commutators

Let us start with the four-dimensional case. Our Lagrangian is

$$\mathcal{L} = \mathcal{L}_G + \mathcal{L}_F \tag{1}$$

L_G is the 't Hooft-Polyakov[2] Lagrangian consisting of an SU(2) gauge field V_μ^a (a=1,2,3) and a scalar field ϕ^a in the vector representation

$$\begin{aligned}\mathcal{L}_G &= -\frac{1}{4} F^a_{\mu\nu} F^{a\mu\nu} + \frac{1}{2}(D_\mu\phi)^a (D^\mu\phi)^a - \lambda(\phi^a\phi^a - f^2)^2 \\ F^a_{\mu\nu} &= \partial_\mu V^a_\nu - \partial_\nu V^a_\mu + e\,\epsilon^{abc} V^b_\mu V^c_\nu \\ (D_\mu\phi)^a &= \partial_\mu\phi^a + e\,\epsilon^{abc} V^b_\mu \phi^c\end{aligned} \tag{2}$$

L_F is

$$\begin{aligned}\mathcal{L}_F &= \bar\psi i \not{D}\psi - m\bar\psi\psi - ig\bar\psi\gamma_5\tau^a\phi^a\psi \\ D_\mu &= \partial_\mu + ieV^a_\mu \tfrac{1}{2}\tau^a\end{aligned} \tag{3}$$

where ψ is a Dirac fermion. L_G is known to yield static magnetic monopole solutions, for which

$$
\begin{aligned}
V_i^a &= \frac{1}{e} \epsilon_{iab} \hat{x}_b V(r) \\
\phi^a &= \hat{x}^a F(r)
\end{aligned}
\tag{4}
$$

where V(r) and F(r) vanish linearly at r = 0 and behave as

$$
\begin{aligned}
V(r) &\rightarrow \frac{1}{r} \\
F(r) &\rightarrow f
\end{aligned}
\tag{5}
$$

for $r \rightarrow \infty$.

The next part follows the discussion in Ref. 3). We define the chiral rotation

$$
U(\theta) = \exp\left[i \int j_{05}^a(x)\, \theta^a(x)\, d^3\vec{x}\right]. \tag{6}
$$

Then

$$
\begin{aligned}
U(\theta)\psi(x)U^{-1}(\theta) &= \exp\left[-\frac{i}{2}\gamma_5 \tau^a \theta^a(x)\right]\psi(x) \\
&\equiv S(x)\psi(x) .
\end{aligned}
\tag{7}
$$

We take θ^a to be

$$
m + ig\gamma_5 \phi^a(x)\tau^a = \rho(x) \exp\left[i\gamma_5 \tau^a \theta^a(x)\right]. \tag{8}
$$

Now

$$
U(\theta)\, \mathcal{L}_F\, U^{-1}(\theta) \equiv \mathcal{L}_F(\theta) = \bar{\psi}(i\partial\!\!\!/ - e B\!\!\!/)\psi - g\bar{\psi}\rho\psi \tag{9}
$$

B includes a vector and an axial vector part

$$
\begin{aligned}
B_k &= S^{-1} V_k S - \frac{i}{e} S^{-1} \partial_k S \\
V_k &\equiv V_k^a \frac{1}{2} \tau^a \\
B_k &\equiv B_k^a \frac{1}{2} \tau^a .
\end{aligned}
\tag{10}
$$

The term proportional to γ_5 gives the axial vector part in B .

$$e B_k^a = \epsilon_{kab} \hat{x}_b \left[V(r) + (1-\cos\theta)\left(\frac{1}{r} - V(r)\right)\right]$$
$$- \gamma_5 \left[\hat{x}_a \partial_k \theta + \sin\theta \left(\delta_{ak} - \hat{x}_a \hat{x}_k\right)\left(\frac{1}{r} - V(r)\right)\right] \quad (11)$$

with $\theta^a = \hat{x}^a \theta$. In the monopole case,

$$B_k^a \longrightarrow \frac{1}{e}\, \epsilon_{kab}\, \hat{x}_b \frac{1}{r} \quad (12)$$

as $r \to \infty$. Thus B_k^a behaves asymptotically like the original vector potential.

The chiral rotation also introduces into the Lagrangian an extra term coming from the change in the Jacobian of the functional integral. That term is a function of θ but does not affect the coupling to fermions. We thus are not concerned with it.

Take now $|M\rangle$ to be the one magnetic monopole state. To discuss the induced fermion current in that state, we need to compute

$$\langle j_\mu \rangle \equiv \langle M | j_\mu(x) | M \rangle_{\mathcal{L}_F} =$$
$$= \langle M | U(\theta)\, j_\mu(x)\, U^{-1}(\theta) | M \rangle_{\mathcal{L}_F(\theta)} . \quad (13)$$

Now

$$U(\theta)\, j_0(x)\, U^{-1}(\theta) = j_0(x) + \frac{e}{4\pi^2} \partial_k \left[H_k^a(x)\, \theta^a(x) \right] \quad (14)$$

as follows from the anomalous commutator[4),5)] (related to the axial anomaly),

$$\left[j_0(\vec{x} t),\, j_{05}^a(\vec{y} t) \right] = \frac{i}{4\pi^2}\, e\, H_k^a(\vec{y} t)\, \partial_k \delta^{(3)}(\vec{x} - \vec{y}) . \quad (15)$$

(Note the extra factor 1/2 as compared with Ref. 5) coming from $\mathrm{Tr}\left[(1/2\tau^a)(1/2\tau^b)\right] = (1/2)\delta^{ab}$, and that our γ_5 has an extra minus.) H_k^a are the magnetic field strengths of V_k^a.

The <u>main</u> point now is that

$$\int d^3\vec{x}\, \langle M | j_0(x) | M \rangle_{\mathcal{L}_F(\theta)} = 0 \quad . \quad (16)$$

This we shall demonstrate in the next section. With Eq. (14) it now follows immediately that

$$Q_F = \frac{e}{4\pi^2} \int d^3\vec{x}\, \partial_k \left[H_k^a(x) \theta^a(x) \right] . \tag{17}$$

Now, dealing with fields H and angles θ regular everywhere, the integral in Eq. (17) can be transformed to a surface integral at infinity only. For the monopole configuration

$$\theta^a(\vec{x}) = \hat{x}^a \theta(r)$$

$$H_k^a(\vec{x}) \underset{r\to\infty}{\longrightarrow} \frac{\hat{x}^a \hat{x}_k}{e r^2}$$

thus obtaining

$$Q_F = \frac{1}{\pi} \theta(\infty) = \frac{1}{\pi} \operatorname{arctg} \frac{gf}{m} \tag{18}$$

which is the result originally obtained by Goldstone and Wilczek[1] from a direct computation.

3. Conserved Currents and Regularization

We first want to prove the statement in Eq. (16). Denote

$$J_\mu(x;\theta) \equiv \langle S | j_\mu(x) | S \rangle_{L_F(\theta)} \tag{19}$$

where $|S\rangle$ is some soliton state, not necessarily the monopole, and $L_F(\theta)$ thus represents an interaction with fermions of external static vector and axial vector currents, and a scalar density $\rho(\vec{x})$.

We assume that a suitable regularization has been made, such that

$$\partial^\mu J_\mu(x;\theta) = 0 \tag{20}$$

for <u>any</u> external field configuration. (After all the Lagrangian $L_F(\theta)$ is invariant under $\psi \to e^{i\alpha}\psi$ for constant α, with any external fields.) We now integrate Eq. (20) over a four-dimensional volume V as in Fig. 1.

We extend our static fields to all times by multiplying with a function $T(t)$ such that $T(-\infty) = 0$ and $T(0) = 1$. We thus obtain

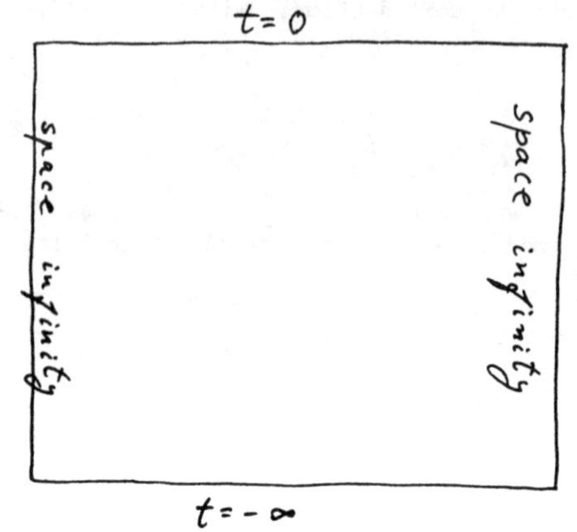

Fig. 1 - Volume V.

$$Q_F + \int_{(\text{space infinity walls})} J_\alpha(x;\theta)\, dS^\alpha = 0 \tag{21}$$

which implies that, for external fields such that $J_\mu(x;\theta)$ is regular at any finite point, the value of Q_F depends only on the behaviour at spacial infinity.

To summarize : for $J_\mu(x;\theta)$ that is conserved for any external field configuration, and chosing the external fields such that J_μ is regular everywhere, the fermionic charge Q_F of the state $|S\rangle$ depends only on the asymptotic behaviour of the external fields.

For the case of the monopole configuration, the external axial source in $L_F(\theta)$ vanishes at infinity faster than $1/r$, while the vector source behaves like in the monopole case ($1/r$ with the same coefficient). There is also a scalar source $\rho(\vec{x})$ that behaves like a constant at infinity. Thus the value of Q_F is governed by asymptotic behaviour of the vector and scalar sources. But for pure vector and scalar sources the induced current J_μ is zero (as follows from G parity). Thus Eq. (16) has been established.

4. Comparison with Explicit Expressions

To get a better understanding of the discussion in the previous section, we are now going to use the explicit expression for the induced fermionic current as given originally in Ref. 1).

Starting from a Lagrangian that involves (scalar, pseudoscalar, vector and axial vector) external sources, the explicit expression for the induced fermionic current is given as [1)]

$$\langle j^{\mu} \rangle = \frac{1}{12\pi^2}\,\epsilon^{\mu\alpha\beta\gamma}\,\epsilon_{dabc}\Big[\frac{\phi_d}{(\phi^2)^2}(D_\alpha\phi)_a(D_\beta\phi)_b(D_\gamma\phi)_c$$

$$+\,\frac{3}{4}\,e\,F_{\alpha\beta,ab}\,\frac{\phi_d}{\phi^2}(D_\gamma\phi)_c\Big] \tag{22}$$

where an O(4) notation has been used for the scalar-pseudoscalar and vector-axial vector combinations $(a,b,c,d = 1,2,3,4)$. Thus the Yukawa couplings are

$$g\,\bar{\psi}\,[\phi_4 + i\gamma_5\,\vec{\tau}\cdot\vec{\phi}]\,\psi \tag{23}$$

while the field strengths $F_{\alpha\beta,ab}$ are constructed from the potentials $A_{\alpha,ab}$ such that

$$A_{\alpha,ab} = \begin{cases} \epsilon_{abc}\,V_{\alpha,c} & a,b,c = 1,2,3 \\ A_{\alpha,a} & a=1,2,3 \quad b=4 \end{cases} \tag{24}$$

V are the vector potentials and A the axial potentials.

Equation (22) yields a conserved induced current for vector external sources, but for general vector and axial vector external sources it does not yield a conserved current[1)]. Thus the regularization employed in Ref. 1) is not the one we use for our J_μ of Eq. (19).

To see the effect more clearly consider first the case of the Lagrangian Eq. (1), which includes a Higgs coupling (with $g\phi_4 = M$ in Eq. (23)) and a pure vector source. This is the case considered originally by Goldstone and Wilczek[1)], and we reproduced their result for Q_F by use of the anomalous commutators without having to compute the induced current explicitly. Suppose we now adopt Eq. (22) for axial as well as vector sources. Since the right-hand side of Eq. (22) is explicitly O(4) invariant, it should hold for the Lagrangian $L_F(\theta)$ as well, since it is obtained from L_F by a chiral rotation. But since numerically the right-hand side of Eq. (22) has not changed, we must get Q_F for the charge also with the interactions as in $L_F(\theta)$, in contradiction with our demonstration of Eq. (16).

What went wrong ? After all, the expression in Eq. (22) remained conserved also after the chiral rotation, namely also for the vector and axial sources in $L_F(\theta)$, since it is numerically identical to the case of the sources in L_F, namely no external axials. The reason is that the expression in Eq. (22) remained conserved only for special axial sources, those obtained by a chiral rotation from the pure vector case as in Eq. (11). For our proof in Section 3 we used a regularization such that J_μ is conserved for <u>any</u> external configuration, which is a different regularization from that of Goldstone and Wilczek [1]. It follows from our discussion in Section 3 that our regularization is the one to be used in conjunction with the anomalous commutators. The expression of Ref. 1) is consistent with our results only for vector external sources.

5. The Two-Dimensional Cases

The method described in Sections 2 and 3 for the four-dimensional case can of course, with some appropriate changes, be applied also in the two-dimensional case. Several issues, however, are much simpler here. The fact that the gauge invariant current carries no charge for $L_F(\theta)$ can here be easily demonstrated explicitly.

We take the Lagrangian

$$\mathcal{L} = i\bar{\psi}\not{\partial}\psi - g\bar{\psi}(\phi_1 + i\gamma_5\phi_2)\psi + \mathcal{L}_1(\phi) \tag{25}$$

L_1 is the Lagrangian of $\phi = (\phi_1, \phi_2)$ which has soliton solutions and ψ is a Dirac fermion. The chiral rotation we employ now is

$$U(\theta) = \exp\left[\frac{i}{2}\int j_{05}(x)\,\theta(x)\,dx\right] \tag{26}$$

with

$$\operatorname{tg}\theta(x) = \frac{\phi_2(x)}{\phi_1(x)} \tag{27}$$

We have

$$\phi_1(x) + i\gamma_5\phi_2(x) = \rho(x)\exp[i\gamma_5\theta(x)]$$

$$U(\theta)\psi(x)U^{-1}(\theta) = \left\{\exp\left[-\frac{i}{2}\gamma_5\theta(x)\right]\right\}\psi(x)$$

and the fermion part L_F is transformed as

$$U(\theta)\mathcal{L}_F U^{-1}(\theta) \equiv \mathcal{L}_F(\theta) = \bar{\psi}i\not{\partial}\psi + \bar{\psi}\gamma_\mu\gamma_5(\partial^\mu\theta)\psi - g\bar{\psi}\rho\psi$$

$L_1(\phi)$ remains invariant. Also here we ignore the extra term from the Jacobian in the functional integral, which in the two-dimensional case is $(1/8\pi)(\partial_\mu\theta)^2$, since it does not contribute to the interaction with the fermions.

Now

$$U(\theta)\, j_0(x)\, U^{-1}(\theta) = j_0(x) + \frac{1}{2\pi}\,\theta'(x) \tag{28}$$

where here the second term comes from the two-dimensional anomalous commutator

$$[j_0(x), j_{05}(y)] = \frac{i}{\pi}\,\delta'(x-y)\ . \tag{29}$$

Let $|S\rangle$ denote the soliton.

$$\begin{aligned} Q_F &= \int dx\, \langle S|\, j_0(x)\, |S\rangle_{\mathcal{L}_F} \equiv \\ &\equiv \int dx\, \langle S|\, U(\theta)\, j_0(x)\, U^{-1}(\theta)\, |S\rangle_{\mathcal{L}_F(\theta)} = \\ &= \frac{1}{2\pi}\int dx\, \theta'(x) = \frac{1}{2\pi}\,[\theta(\infty) - \theta(-\infty)] \end{aligned} \tag{30}$$

where, as in the four-dimensional case

$$\int dx\, \langle S|\, j_0(x)\, |S\rangle_{\mathcal{L}_F(\theta)} = 0\ . \tag{31}$$

Here we can actually demonstrate this explicitly. The interaction in $L_F(\theta)$ is that of a vector potential $V_\mu = \varepsilon_{\mu\nu}\partial^\nu\theta$ and a scalar source ρ. The explicit demonstration of Eq. (31) is immediate[6].

Another curious fact should be mentioned. When constructing the gauge invariant current under the interaction $L_F(\theta)$, we employ a point splitting procedure and an exponential of a line integral of the vector potential V_μ. The original current was constructed under the sources in L_F, and had no vector sources. It turns out[6] that the contribution of the point splitting (which after the chiral rotation amounts to θ at different points) and that of the addition of the line integral are equal, each one yielding $(1/4\pi)[\theta(\infty)-\theta(-\infty)]$ to Q_F.

Finally, when $\theta(\infty) = -\theta(-\infty) = \theta$ we get

$$Q_F = \frac{1}{\pi}\,\Theta = \frac{1}{\pi}\,\mathrm{arc\,tg}\,\frac{gf}{m} \tag{32}$$

where we took $g\phi_1 \equiv m$ and $g\phi_2 \to \pm f$ as $x \to \pm\infty$. The two-dimensional result, Eq. (32), is identical to the four-dimensional result, Eq. (18).

References

1) J. Goldstone and F. Wilczek, Phys.Rev.Lett. 47 (1981) 988.

2) G. 't Hooft, Nucl.Phys. B79 (1974) 276;
A.M. Polyakov, JETP Lett. (1974) 194.

3) Y. Frishman, D. Gepner and S. Yankielowicz, "Fractional Charge and Anomalous Commutators", Phys.Lett. B (in print).

4) R. Jackiw and K. Johnson, Phys.Rev. 182 (1969) 1459.

5) S.L. Adler, in "Lectures in Elementary Particles and Quantum Field Theory", Brandeis University Summer Institute (1970).

6) W.A. Bardeen, S. Elitzur, Y. Frishman and E. Rabinovici, Nucl.Phys. B218 (1983) 445.

NON-LOCAL CHARGES AND MONODROMY MATRICES IN INTEGRABLE FIELD THEORIES

H.J. de Vega
LPTHE, Universitè P.et M.Curie, Paris

ABSTRACT

The non-local charges and their generating functionals (monodromy matrices) are investigated for two-dimensional field theories like sigma models and chiral invariant fermion models. Their exact quantum matrix elements are derived as well as the classical (Poisson brackets) and quantum algebra. Both classical and quantum algebras of monodromy matrices turn to be bilinear Lie algebras. Classical and quantum R-matrices are given.

Classical and quantum mechanical systems possessing as many commuting (and conserved) quantities as degrees of freedom are called integrable. Then, field theories and statistical models are called integrable provided they have an infinite number of conserved quantities.

The classical and quantum inverse scattering methods permit the construction of such integrable theories.

The classical inverse scattering methods (CISM) can be formulated as follows[1]. Let L and M be non-singular matrices that depend on a field $\phi(t,x)$ and a spectral parameter λ (ϕ may be a multicomponent object). The linear system

$$\frac{\partial\psi}{\partial x} = L(\lambda, \phi(t,x))\,\psi(t,x) \tag{1}$$

$$\frac{\partial\psi}{\partial t} = M(\lambda, \phi(t,x))\,\psi(t,x) \tag{2}$$

will not be in general compatible. A sufficient condition of compatibility is the vanishing of the commutator $(\partial_x - L\ ,\ \partial_t - M)$ i.e.

$$\frac{\partial L}{\partial t} - \frac{\partial M}{\partial x} + [L, M] = 0 \quad . \tag{3}$$

This zero-curvature condition must be identically fulfilled in λ and gives for $\phi(t,x)$ a non-linear partial differential equation (PDE) or a system of PDE. This is the PDE or system of PDE that one can solve with the help of the linear system (1). So, the CISM can only be applied to PDE admitting a representation like eq. (3). This has been done for a restricted, although wide, class of equations.

The monodromy matrix plays here an important role. It is essentially the "S-matrix" of eq. (1) considered as a one-dimensional scattering problem with $\phi(t,x)$ as the potential and t as an extra parameter (not the time for the scattering process). Let us assume boundary conditions for $\phi(t,x)$ such that the limits

$$\lim_{x \to \pm\infty} L(\lambda, \phi(t,x)) = L_\pm(\lambda) \tag{4}$$

exist and are finite. We consider matrix solutions $\psi_\pm(t,x;\lambda)$ of eq. (1) such that

$$\psi_\pm(t,x;\lambda) = e^{x L_\pm(\lambda)} \qquad (x \to \pm\infty) \tag{5}$$

So, for $x \to \pm\infty$, one has

$$\psi_\mp = e^{x L_\pm(\lambda)} T(\lambda, t) \tag{6}$$

$T(\lambda,t)$ is the so-called monodromy matrix. The clue of the CISM lies in the fact that eq. (2) implies a harmonic time dependence for $T(\lambda,t)$.

There exists a large class of relativistic field theories admitting an associate linear system of the type of eqs. (1) - (2): root system fields[2)], the complex-sine-Gordon theory[3)], generalized sigma models[4] and fermionic theories with quartic interactions[5,6]. These theories admit supersymmetric extensions which are also integrable.

A particularly interesting class of integrable theories consists in the generalized sigma models, the Gross-Neveu model and the generalized chiral models of ref. (6). All these theories are classically conformal invariant. As quantum theories, they are renormalizable models exhibiting asymptotic freedom and dynamical mass generation through dimensional transmutation. The equations of motion writes

$$\partial_\mu A^\mu(x) = 0 \tag{7}$$

$$\partial_0 A_1 - \partial_1 A_0 + [A_0, A_1] = 0 \ . \tag{8}$$

For sigma models

$$A_\mu(x) = g^{-1}\partial_\mu g(x) \tag{9}$$

where g takes values in a symmetric space G/H. G stands for a Lie group (compact or not) and HCG is an invariant subgroup under an involutive automorphism. For G = SU(n) or O(N), H = ϕ, we get the chiral models. When G = O (n+m), H = O(n) x O(m), we get gauge invariant models. This last example gives for n = 1, the O(m+1) invariant non-linear sigma model. For all these sigma models, the Lagrangian reads

$$\mathcal{L} = \mathrm{Tr}\left[\partial_\mu g \ \partial_\mu g^{-1}\right] \tag{10}$$

for chiral invariant fermionic models[6]

$$A_\mu(x) = -4ig\sum_\alpha t_\alpha(\bar{\Psi}\gamma_\mu t^\alpha \psi)$$

there t_α are the generators of some group G

$$[t_\alpha, t_\beta] = f^\gamma{}_{\alpha\beta}\, t_\gamma , \qquad t^\alpha = k^{\alpha\beta} t_\beta ,$$

where $k_{\alpha\beta} = (k^{-1})^{\alpha\beta}$ is the Killing form of G. $\psi_{a\alpha}$ (a = 1, ..., N; α = 1, 2) is a fermionic field.

The Lagrangian reads

$$\mathcal{L} = i\bar{\psi}\not{\partial}\psi + g(\bar{\psi}\gamma_\mu t_\alpha \psi)(\bar{\psi}\gamma^\mu t^\alpha \psi) . \tag{11}$$

We deal with both commuting (C) and anti-commuting (AC) spinors.

The main point is that eqs. (7) and (8) admit the associated lines system[4)]

$$\left(\partial_\mu - \lambda\, \epsilon_{\mu\nu}(\partial^\nu + A^\nu)\right)\psi(x,\lambda) = 0, \tag{12}$$

where $\epsilon_{\mu\nu}$ is the antisymmetric tensor with $E_{01} = +1$. Eq. (12) is invariant under the transformation

$$g(x) \to g(x)^{-1}, \quad \lambda \to \lambda^{-1}; \quad \psi \to g\,\psi\, g^{-1} .$$

A family of conserved currents can be introduced as

$$J_\mu(x,\lambda) = \epsilon_{\mu\nu}\, \partial^\nu \psi_-(x,\lambda) , \tag{13}$$

where ψ_- obeys eqs. (5), (6) and (12). We assume the finite energy boundary conditions

$$\lim_{|x|\to\infty} A_\mu(x,t) = 0 . \tag{14}$$

Then $L_+ = L_- = 0$ and $T(\lambda)$ turns to be conserved. The charges associated to the current (13) are time-independent

$$\int_{-\infty}^{\infty} dx\, J_0(x,t) = T(\lambda) - 1 . \tag{15}$$

Expanding in powers of λ, one gets an infinite number of non-local conserved charges[4)]

$$\log T(\lambda) = \sum_{n=0}^{\infty} \lambda^{n+1} Q_n \tag{16}$$

$$Q_0 = \int dx\, A_0 , \quad Q_1 = -\frac{1}{2}\int dx\,dy\, \epsilon(x-y) A_0(x) A_0(y) - \int dx\, A_1 , \tag{17}$$

$$\ldots$$

The absence of a mass scale in the classical theory presents the linear system (12) to provide here angle variables. Only conserved quantities follow from it. For later use, let us give now the transformation of $T(\lambda)$ under parity (P) and time reversal (τ)

$$P: \begin{array}{l} A_\mu(t,x) \to A^\mu(t,-x) \\ T(\lambda) \to T(-\lambda)^{-1} \end{array} ; \quad P\tau: \begin{array}{l} A_\mu(t,x) \to -A_\mu(-t,-x) \\ T(\lambda) \to T(\lambda)^{-1} \end{array} . \tag{18}$$

There exists an infinite dimensional Lie algebra of symmetry transformations in the space of solutions of eqs. (7) and (8) like for example in the case of the chiral sigma model. It has the structure of a loop algebra[7)]. For this example, where g takes values in a group G with Lie algebra ξ their infinitesimal symmetry transformations read

$$\begin{aligned} g(x) &\to g(x) - \epsilon M_a^{(\pm)\lambda}(g) , \quad \epsilon \ll 1 \\ M_a^{(\pm)\lambda} &\equiv g V_a^{(\pm)}(x,\lambda) \\ V_a^{(\pm)}(x,\lambda) &= \psi_\pm(x,\lambda)^{\pm 1}\, t_a\, \psi_\pm(x,\lambda)^{\mp 1} \end{aligned} \tag{19}$$

where the $\{t_a\}$ are a basis of ξ,

$$[t_a, t_b] = f_{abc} t_c .$$

If $g(x)$ is a solution of eqs. (7) and (8), then $g - \epsilon M g$ is a solution too up to $O(\epsilon^2)$. Moreover

$$[M_a^{(\pm)\lambda}, M_b^{(\pm)\mu}] = \frac{f_{abc}}{\lambda-\mu}[\lambda M_c^{(\pm)\lambda} - \mu M_c^{(\pm)\mu}] \tag{20}$$

and expanding in powers of λ and μ

$$M_a^{(\pm)\lambda} = \sum_{n=0}^{\infty} M_a^{(\pm)n} \lambda^n \tag{21}$$

$$[M_a^{(\pm)n}, M_b^{(\pm)m}] = f_{abc} M_c^{(\pm)n+m} . \tag{22}$$

We have two isomorphic loop algebras $\{M^{(+),n}\}$ and $\{M^{(-1),n}\}$ which do not commute with each other. It must be noted that the transformations (19) are not canonical since they do not preserve the Poisson brackets[8,9]. One can also consider the canonical transformations generated by the non-local charges Q_n or by $T(\lambda)$ itself. They are related to the $M^{(\pm),n}$. For example Q_1 generates $M^{(+),1}+M^{(-),1}$.

An important problem is to determine the canonical algebra (Poisson brackets) of monodromy matrices and non-local charges from the basic Poisson brackets. For chiral sigma models we have

$$\{g(t,x) \overset{\otimes}{,} \pi(t,y)\} = \delta(x-y)P , \tag{23}$$

where $\pi = \partial_0(g^{-1})^T$; $P_{ab,cd} = \delta_{ad}\delta_{bc}$ and we use the tensor product notation

$$(A \otimes B)_{ab,cd} = A_{ac} B_{bd} .$$

For fermionic models with commuting spinors, we have

$$\{\psi_{a\alpha}(t,x), \psi_{b\beta}(t,y)\} = i\delta_{ab}\delta_{\alpha\beta}\,\delta(x-y) . \tag{24}$$

If $\psi_{a\alpha}$ is a Grassmann variable, we use

$$\{A,B\} = i\int dx \sum_{a\alpha} A\left(\frac{\overleftarrow{\delta}}{\delta\psi^{\dagger}_{a\alpha}}\frac{\overrightarrow{\delta}}{\delta\psi_{a\alpha}} + \frac{\overleftarrow{\delta}}{\delta\psi_{a\alpha}}\frac{\overrightarrow{\delta}}{\delta\psi^{\dagger}_{a\alpha}}\right)B \,. \tag{25}$$

The field dependence of $T(\lambda)$ is defined by the linear system (12)

$$\partial_x \psi_- = L\psi_- \,, \quad L = \frac{\lambda A_0 + A_1}{\lambda^2 - 1} \,. \tag{26}$$

Then the P.B. of monodromy matrices follows from the classical current algebra:

$$\{T(\lambda) \overset{\otimes}{,} T(\mu)\} = \int dx dy \frac{\delta T(\lambda)}{\delta L(x,\lambda)_{ab}} \otimes \frac{\delta T(\mu)}{\delta L(y,\mu)_{cd}} \{L(x,\lambda) \overset{\otimes}{,} L(y,\mu)\}_{ac\,bd} \,. \tag{27}$$

One gets from eqs. (23) and (27) for the chiral sigma model

$$\{A_0(x) \overset{\otimes}{,} A_0(y)\} = \delta(x-y)[1 \otimes A_0, P] \,, \quad \{A_1(x) \overset{\otimes}{,} A_1(y)\} = 0 \,, \tag{28}$$

$$\{A_1(x) \overset{\otimes}{,} A_0(y)\} = \delta(x-y)[1 \otimes A_1, P] - \partial_x\{\delta(x-y)\, g^{-1}(x)\, g(y) \otimes 1\}P \,.$$

It follows from eq. (24) or (25) and eq. (27) for fermionic models

$$\{A^{\mu}(x) \overset{\otimes}{,} A^{\nu}(y)\} = 4g\delta(x-y)[1 \otimes A^{|\mu-\nu|}(x), \Pi] \,. \tag{29}$$

Π is a constant matrix that depends on the model chosen. One has for both C and AC chiral models

$$\Pi = \sum_{\alpha=1}^{\dim \mathcal{G}} t_{\alpha} \otimes t^{\alpha} \,. \tag{30}$$

In particular $\pi = P$ for $\xi = SU(N)$.

For the C-Gross-Neveu and AC-Gross-Neveu we have

$$\Pi = \frac{1}{2}(\Sigma + P) \ , \quad \Pi = \frac{1}{2}(\Delta - P), \tag{31}$$

respectively. Here $\Sigma_{ac,bd} = \epsilon^{ac}\epsilon_{bd}$, $\epsilon = \begin{pmatrix} 0 & I_N \\ -I_N & 0 \end{pmatrix}$ and $\Delta_{ac,bd} = \delta_{ac}\delta_{bd}$.

The integrand in eq. (27) turns to be a total derivative. So, one gets for fermions[6)]

$$\{T(\lambda) \overset{\otimes}{,} T(\mu)\} = \frac{4g}{\lambda - \mu}\,[\Pi, T(\lambda) \otimes T(\mu)] \ . \tag{32}$$

For chiral sigma models, a detailed analysis[9)] shows that this P.B. is not uniquely defined. Moreover, no definition of the P.B. consistent with the basic properties of Lie Algebras (antisymmetry and Jacobi) has been found. This disease can be traced back to the non-ultralocal character of the current algebra (28) (derivitives of Dirac deltas). The non-ultralocality does not seem to cause difficulties in the quantum theory.

In conclusion, the canonical algebra of monodromy matrices for fermionic models is a quadratic Lie Algebra. The P.B. of two T's is not a linear but a quadratic expression in the T's.

Inserting the expansion

$$T(\lambda) = \sum_{n=0}^{\infty} \lambda^{-n} T_{n-1} \ , \quad T_{-1} = 1 \tag{33}$$

in eq. (32), one gets the algebra

$$\begin{aligned} \{T_n \overset{\otimes}{,} T_0\} &= 4g\,[\Pi, T_n \otimes 1] \\ \{T_1 \overset{\otimes}{,} T_1\} &= 4g\,[\Pi, T_2 \otimes 1 + T_1 \otimes T_0], \ldots \end{aligned} \tag{34}$$

In general

$$\{T_n \overset{\otimes}{,} T_m\} = 4g\left[\Pi, \sum_{l=0}^{m} T_{l+n} \otimes T_{m-l-1}\right] . \qquad (35)$$

We see that the highest charge on the r.h.s. is T_{n+m}, but unlike for the case of a loop algebra (eq. (22)), a bunch of bilinear terms are present.

Let us now discuss the quantum algebra of currents and monodromy matrices. The current algebra for fermionic chiral fields reads

$$[A_\mu(x) \overset{\otimes}{,} A_\nu(y)] = 4ig\,\delta(x-y)[\mathbb{1} \otimes A_{|\mu-\nu|}, \Pi] + \frac{8ig^2}{\pi}\delta'(x-y)\,\Pi\,\delta_{1,|\mu-\nu|} .$$

This last term is the so-called Schwinger term. It is finite in two space-time dimensions. The coefficient has been computed assuming the fermion field to be free[6]. Consequently

$$[L(x,\lambda) \overset{\otimes}{,} L(y,\mu)] = -\frac{4ig}{\lambda-\mu}\delta(x-y)[\Pi, L(x,\lambda)\otimes \mathbb{1} + \mathbb{1}\times L(y,\mu)] - \frac{\lambda+\mu}{(\lambda^2-1)(\mu^2-1)}\frac{8g^2 i}{\pi}\Pi\,\delta'(x-y) . \qquad (36)$$

In Fourier space

$$L(x,\lambda) = \int \frac{dk}{\pi} e^{ikx}\tilde{L}(k,\lambda)$$

and

$$[\tilde{L}(k,\lambda) \overset{\otimes}{,} L(k',\mu)] = -\frac{2ig}{\lambda-\mu}[\Pi, \tilde{L}(k+k',\lambda)\otimes \mathbb{1} + \mathbb{1}\otimes L(k+k',\mu)] + \frac{\lambda+\mu}{(\lambda^2-1)(\mu^2-1)} 4g^2\,\Pi\, k\,\delta(k+k') . \qquad (37)$$

This relation has the same structure as the commutation relations of a Kac-Moody algebra.

The quantum monodromy matrix is explicitly determined from the following assumptions[9].

a) $T(\lambda)$ exists as a quantum operator and it is conserved.

b) $T(\lambda)$ fulfills the factorization principle.

c) P, τ and internal invariances hold in the quantum theory.

As it is known, absence of particle production and S-matrix factorizability follow from a)[10]. Moreover, a) and c) determine the S-matrix up to CDD poles[11]. Statement b) easily follows in the classical theory from eq. (26). Let the field $g(x)$ form two separated lumps $g(x) = g_1(x)$ for $x \leq a$, $g(x) = \text{const.}$ for $a \leq x \leq b$ and $g(x) = g_2(x)$ for $x \geq b$. Then

$$T_{ab}(\lambda, g(\cdot)) = T_{ac}(\lambda g_1(\cdot))\, T_{cb}(\lambda, g_2(\cdot)) \quad . \tag{38}$$

In ref. (12), a quantum version of eq. (38) has been proposed.

$$T_{ab}(\lambda)\,|\theta_1 c_1, \dots, \theta_k c_k\rangle_{out} = \sum_{\{a_i\}} T_{aa_1}|\theta_1 c_1\rangle \cdots T_{a_{k-1}b}\,|\theta_k c_k\rangle \tag{39}$$

$$T_{ab}(\lambda)\,|\theta_1 c_1, \dots, \theta_k c_k\rangle_{in} = \sum T_{a a_{k-1}}|\theta_k c_k\rangle \cdots T_{a_1 b}\,|\theta_1 c_1\rangle \ .$$

Here θ_i, C_i stand for rapidity and internal quantum numbers of the asymptotic particles ($\theta_i > \theta_j$ for $i > j$). For the non-linear sigma model relativistic and O(N) invariance imply

$$\langle \theta d|T_{ab}(\lambda)|\theta' c\rangle = \delta(\theta-\theta')\big[\delta_{ad}\delta_{bc} f_1(\lambda\theta) + \delta_{ab}\delta_{cd} f_2(\lambda,\theta) + \delta_{ac}\delta_{bd} f_3(\lambda,\theta)\big] \tag{40}$$

where we have also used the time-independence of $T(\lambda)$ and a, b, c, d run from 1 to N.

Since the S-matrix relates in and out states

$$\langle in|S\,T_{ab}(\lambda)|in\rangle = \langle out|T_{ab}(\lambda)\,S|out\rangle \ .$$

For two-particle states, this equation gives a set of relations for the $f_i(\lambda,0)$ that reduces to[9]

$$f_3(\lambda,\theta) = -\frac{2\pi i}{N-2}\frac{f_2(\lambda,\theta)}{\theta+\gamma(\lambda)}\,, \quad f_1(\lambda,\theta) = \frac{2\pi i}{N-2}\frac{f_2(\lambda,\theta)}{\theta+\gamma(\lambda)-i\pi} \tag{41}$$

where γ depends only on λ. The quantum analogue of eq. (18) for P and T invariance gives[9)]

$$\gamma(-\lambda) = -\gamma(\lambda)\,, \quad \gamma(\lambda) = \gamma(\lambda^*)^* \tag{42}$$
$$f_2(\lambda,\theta) = f_2(-\lambda^*,-\theta^*)^*, \quad |f_2|^{-2} = 1+\left(\frac{2\pi}{(N-2)(\theta+\gamma)}\right)^2 .$$

The general solution reads

$$\langle\theta d|T_{ab}(\lambda)|\theta' c\rangle = \delta(\theta-\theta')\,S_{ad,bc}(\theta+\gamma(\lambda))\,e^{i\phi} \tag{43}$$

where S is the two-body S-matrix and $\phi(\lambda,\theta)$ is real for real λ,θ and

$$\phi(-\lambda,-\theta) = -\phi(\lambda,\theta)\,.$$

More explicitly

$$f_i(\lambda,\theta) = \sigma_i(\theta+\gamma(\lambda))\,e^{i\phi(\lambda,\theta)} \tag{44}$$

where

$$\sigma_1 = \frac{2\pi i}{N-2}\frac{\sigma_2}{\theta-i\pi}\,, \quad \sigma_3 = -\frac{2\pi i}{N-2}\frac{\sigma_2}{\theta}$$

$$\sigma_2 = \frac{\Gamma(\Delta+z)\,\Gamma(z+\frac{1}{2})\,\Gamma(\Delta+\frac{1}{2}-z)\,\Gamma(1-z)}{\Gamma(z)\,\Gamma(\Delta+\frac{1}{2}+z)\,\Gamma(\frac{1}{2}-z)\,\Gamma(\Delta+1-z)}\,, \tag{45}$$

$$\Delta = (N-2)^{-1}\,, \quad z = \theta/2\pi i.$$

These results have been checked with the renormalized expression of $T(\lambda)$[10),9,13)]

$$\log T(\lambda) = \lambda \int A_0 dx - \frac{\lambda^2}{2} \int_{|x-y|>\delta} dx\,dy\ \epsilon(x-y) A_0(x) A_0(y)$$

$$- \lambda^2 Z(\delta) \int A_1(x) dx + O(\lambda^3), \tag{46}$$

where $Z(\delta) = \frac{N-2}{2\pi} \ln m\delta$

for the non-linear σ-model and

$$Z(\delta) = \frac{2gN}{\pi} \log m\delta$$

for the chiral SU(N) Gross-Neveu model and $\delta \to 0^+$.

One gets from here for the N L σ m:

$$\langle \theta d | T_{ab}(\lambda) | \theta' c \rangle = \delta(\theta - \theta') \{ \delta_{ab} \delta_{cd} (1 - 2\lambda^2) - \delta_{ac} \delta_{bd}\, 2i\lambda$$
$$\times \left(1 - \frac{\lambda}{\pi}(N-2)\theta\right) + \delta_{ad} \delta_{bc}\, 2i\lambda \left(1 + \frac{\lambda}{\pi}(N-2)(\pi - \theta)\right) + O(\lambda^3) \} \tag{47}$$

and analogous expressions for fermionic models. It follows from eqs. (40) - (47) that the factorization principle holds, the one-particle matrix elements match and

$$\phi(\lambda,\theta) = 0 + O(\lambda^3) \quad (\text{NL}\sigma\text{m}), \qquad \phi(\lambda,\theta) = \pm \frac{\pi}{N} + O(\lambda^3) \quad (\text{SU(N)CGN}) \tag{48}$$

$$\gamma(\lambda) = \frac{\pi}{N-2} \lambda^{-1} + O(\lambda), \qquad \gamma(\lambda) = \frac{\pi}{2gN} \lambda^{-1} + O(\lambda). \tag{49}$$

This suggests that $\phi(\lambda,\theta)$ vanishes identically for the n ℓ σ m.

The algebra of the quantum $T(\lambda)$ is derived as follows. A K-particle matrix element of $T(\lambda)$ reads

$$T_{a\{c'\},b\{c\}} = \langle \theta'_1 c'_1, \ldots, \theta'_K c'_K \text{ out} | T_{ab}(\lambda) | \theta_1 c_1, \ldots, \theta_K c_K \text{ out} \rangle$$

$$= \prod_{j=1}^{K} \delta(\theta_j - \theta'_j) \sum_{\{a_i\}} S_{ac'_1\, a_1 c_1}(\theta_1 + \gamma(\lambda)) \cdots S_{a_{K-1} c'_K, b c_K}(\theta_K + \gamma(\lambda)) . \tag{50}$$

This matrix element is formally identical to the transfer operator in an inhomogeneous two-dimensional vertex model if we make the identifications[14)]

a = particle state $\leftrightarrow$ link state

2 body S-matrix $S_{ab,cd}(\theta k + \gamma(\lambda))$ $\leftrightarrow$ statistical weight

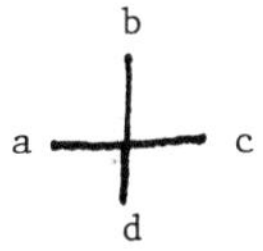

K = number of particles $\leftrightarrow$ number of sites.

One finds from the factorization equations for the S-matrix that $T_{a\{c'\},b\{c\}}$ satisfies

$$R(\lambda,\lambda')\left[T(\lambda) \otimes T(\lambda')\right] = \left[T(\lambda') \otimes T(\lambda)\right] R(\lambda,\lambda'), \tag{51}$$

where

$$R(\lambda,\lambda') = P\, S(\gamma(\lambda) - \gamma(\lambda')). \tag{52}$$

So, we find that the quantum R-matrix for the non-linear sigma model and chiral Gross-Neveu models equals P times the respective two-body S-matrix at $\theta = \gamma(\lambda) - \gamma(\lambda')$. It can be noted that the phases $\phi(\theta_K, \lambda)$ drop in eq. (51). Insertion of eq. (40) - (44) in eq. (51) leads to the result for the O(N) non-linear σ model

$$[T_{ac}(\lambda),T_{bd}(\lambda')]=\frac{2\pi i}{N-2}\left(\frac{T_{bc}(\lambda)T_{ad}(\lambda')-T_{bc}(\lambda')T_{ad}(\lambda)}{\gamma(\lambda)-\gamma(\lambda')}\right.$$

$$\left.+\frac{\delta_{cd}T_{be}(\lambda')T_{ae}(\lambda)-\delta_{ab}T_{ec}(\lambda)T_{ed}(\lambda')}{\gamma(\lambda)-\gamma(\lambda')-i\pi}\right) \tag{53}$$

and for the SU(N) chiral Gross-Neveu model

$$[T_{\alpha\gamma}(\lambda),T_{\beta\delta}(\lambda')]=\frac{2\pi i}{N}\;\frac{T_{\beta\gamma}(\lambda)T_{\alpha\delta}(\lambda')-T_{\beta\gamma}(\lambda')T_{\alpha\delta}(\lambda)}{\gamma(\lambda)-\gamma(\lambda')}\;. \tag{54}$$

In conclusion, the quantum monodromy matrices satisfy a quadratic algebra. As in the classical case (31. (32)), the r.h.s. of eqs. (53) - (54) is bilinear in the T-s and not just linear as for Lie algebras. Quadratic algebras of the type (51) are characteristic of integrable field theories and integrable statistical models[15]. However, all the $T_{ab}(\lambda)$ are conserved here and not just the trace $\sum_a T_{aa}$ as in previous examples[15].

I would like to thank H. Eichenherr and J.M. Maillet for useful discussions.

References

1. See for reviews:
   A.C. Scott, F.Y. Chu and D.W. McLaughlin, Proc. IEEE 61, 1443 (1973);
   M.J. Ablowitz and H. Segur, Solitons and Inverse Scattering Transformations, SIAM Philadelphia (1981);
   L.D. Faddeev, Les Houches Lectures 1982, Saclay preprint T-82-76.
2. A.V. Mikhailov, JETP Lett. 30, 414 (1980);
   S.A. Bulgadaev, Phys. Lett. 96B, 151 (1980);
   O. Babelon, H.J. de Vega and C.M. Viallet, Nucl. Phys. B190, (FS3) 542 (1981).
3. F. Lund and T. Regge, Phys. Rev. D14, 1524 (1976);

3. continued.
K. Pohlmeyer, C.M.P. 46, 207 (1976);
H.J. de Vega and J.M. Maillet, Phys. Lett. 101B, 302 (1981) and LPTHE preprint 82/15;
J.M. Maillet, Phys. Rev. D

4. E. Brézin, C. Itzykson, J. Zinn-Justin and J.B. Zuber, Phys. Lett. 82B, 442 (1979);
H.J. de Vega, Phys. Lett. 87B, 233 (1979);
H. Eichenherr and M. Forger, Nucl. Phys. B155, 381 (1979);
V.E. Zakharov and A.V. Mikhailov, JETP 47, 1017 (1978);
M. Lüscher and K. Pohlmeyer, Nucl Phys. B137 46 (1978).

5. D.J. Gross and A. Neveu, Phys. Rev. D10, 3235 (1974).

6. H.J. de Vega, H. Eichenherr and J.M. Maillet, LPTHE Paris preprint 83/17, June 1983.

7. L. Dolan, Phys. Rev. Lett. 47, 1371 (1981);
K. Veno, Kyoto preprint RIMS-374 (1981);
H. Eichenherr in Lecture Notes in Physics, vol. 180, Springer.

8. M.C. Davies, P.J. Houston, J.M. Leinaas and A.J. MacFarlane, Phys. Lett. 119B, 187 (1982).

9. H.J. de Vega, H. Eichenherr and J.M. Maillet, LPTHE Paris preprint 83/9, March 1983.

10. M. Lüscher, Nucl. Phys. B135, 9 (1978).

11. See for a review, A.B. Zamolodchikov and Al. B. Zamolodchikov, Ann. Phys. 80, 253 (1979).

12. Al. B. Zamolodchikov, Dubna preprint EL-11485 (1978).

13. H.J. de Vega, H. Eichenherr and J.M. Maillet (in preparation).

14. See for example:
O. Babelon, H.J. de Vega and C.M. Viallet, Nucl. Phys. B200, (FS4), 266 (1982).

15. See for reviews: the third reference under (1) and
P.P. Kulish and E.K. Sklyanin, in Lecture Notes in Physics, Vol 151, Springer Verlag, February, 1982.

STOCHASTIC QUANTIZATION

B. Sakita*
Physics Department, City College of the City University of New York,
New York, New York 10031

The stochastic quantization of Parisi and Wu is discussed by emphasizing the basics required for this method. The short remarks are also given for the subjects of more recent developments such as Langevin equation for lattice gauge theory, stochastic quantization of Fermi fields and supersymmetric field theories, and the stochastic regulariztion.

*Supported by National Science Foundation grant No. PHY-82-15364 and PSC-BHE Faculty Research Award No. RF-6-63264.

Last summer I gave two talks on the same subject together with the application to the Eguchi-Kawai type large N reduction in conferences in Japan. A large portion of the actual presentation to the present workshop was not very much different from the written reports published in these proceedings.[1,2]

In the present report, therefore, I will emphasize only the basics required for the Parisi-Wu stochastic quantization.[3]

Euclidean Green's Function

Let $\phi_\ell(x)$ be a real Bose field, where x is a point in d-dimensional space time and let ℓ be a set of internal and external (Lorentz) indices. The Euclidean Green's functions are defined by

$$\begin{aligned} &< \phi_{\ell_1}(x_1)\, \phi_{\ell_2}(x_2) \ldots\ldots \phi_{\ell_n}(x_n) > \\ &= \frac{\int D\phi \;\; \phi_{\ell_1}(x_1)\;\; \phi_{\ell_2}(x_2)\ldots\ldots \phi_{\ell_n}(x_n)\; e^{-S[\phi]}}{\int D\phi \;\; e^{-S[\phi]}} \end{aligned} \tag{1}$$

where $S[\phi]$ is an action.

Fokker-Planck Equation

One considers as an equilibrium limit for a system governed by Fokker-Plank equation:

$$- \frac{\partial}{\partial\tau} P\,[\phi,\, \tau] = H_{FP}\, P[\phi, \tau] \tag{2}$$

Namely, the probability distribution $P[\phi, \tau]$ approaches to the Boltzman distribution:

$$P[\phi, \tau] \underset{\tau\to\infty}{\to} e^{-S[\phi]} \Big/ \int \phi\, D\, e^{-S[\phi]} \tag{3}$$

Fokker-Planck Hamiltonian

What are the possible forms of H_{FP}?

Being a probability distribution $P[\phi,\tau]$ is normalized as

$$\int D\phi\, P[\phi,\tau] = 1. \tag{4}$$

Taking time derivative and using (2) one obtains

$$\int D\, \phi\, H_{FP}\, P[\phi,\tau] = 0. \tag{5}$$

This is satisfied if H_{FP} contains a functional derivative operator on its left.

Let E_n be an eigenvalue of H_{FP} and χ_n be the corresponding eigenfunction:

$$H_{FP}\, \chi_n [\phi] = E_n\, \chi_n [\phi] \tag{6}$$

The solution of Fokker-Planck equation is then given by

$$P[\phi,\tau] = \sum_n e^{-E_n \tau} C_n\, \chi_n [\phi] \tag{7}$$

If

$$E_n > 0 \qquad (n \neq 0)$$

and (8)

$$E_0 = 0 ,$$

namel, there is an energy gap above the ground state, and further the ground state is non-degenerate, one obtains

$$P[\phi,\tau] \underset{\tau\to\infty}{\to} \chi_0 [\phi] \tag{9}$$

Comparing (9) with (3) one requires

$$\chi_0[\phi] = e^{-S[\phi]}/\int D\phi e^{-S[\phi]} \tag{10}$$

Since $\chi_0 [\phi]$ satisfies

$$\left(\frac{\delta}{\delta\phi_\ell(x)} + \frac{\delta S[\phi]}{\delta\phi_\ell(x)}\right) \chi_0[\phi] = 0 \tag{11}$$

and since $\chi_0[\phi]$ is the zero energy eigenfunction, one requires the form of H_{FP} such that it contains the operator $\left(\frac{\delta}{\delta\phi_\ell} + \frac{\delta S}{\delta\phi_\ell}\right)$ on its right. Thus, one obtains

$$H_{FP} = \sum_{\ell\ell'} \iint dxdx' \frac{\delta}{\delta\phi_\ell(x)} K_{\ell\ell'} (x,x';\phi) \left(\frac{\delta}{\delta\phi_{\ell'}(x')} + \frac{\delta S}{\delta\phi_{\ell'}(x')}\right) \tag{12}$$

In this stage $K_{\ell\ell'}(x,x';\phi)$ is arbitrary, but we shall assume it as a functional of $\phi_\ell(x)$ only (i.e. not of $\frac{\delta}{\delta\phi_\ell(x)}$).

In order to make H_{FP} manifestly Hermitean we perform a similarity transformation defined by

$$\hat{H}_{FP} = e^{1/2\, S[\phi]} H_{FP}\, e^{-1/2\, S[\phi]} \tag{13}$$

Then, we obtain

$$\hat{H}_{FP} = \sum_{\ell\ell'} \iint dx dx' \, Q_\ell^\dagger(x) \, K_{\ell\ell'}(x,x';\phi) \, Q_{\ell'}(x') \tag{14}$$

where

$$Q_\ell(x) = -i \left(\frac{\delta}{\delta\phi_\ell(x)} + \frac{\delta S}{\delta\phi_\ell(x)}\right) \tag{15}$$

Thus if K is Hermitean then $\hat{H}_{FP}$ is Hermitean. Further we assume K is written as

$$K_{\ell\ell'}(x,y;\phi) = \sum_m G_{\ell m}(x;\phi) \, G_{\ell' m}(y;\phi) \tag{16}$$

where G is a real functional. $\hat{H}_{FP}$ is a sum of the form of $a^\dagger a$ so that the eigenvalues of $\hat{H}_{FP}$ accordingly H_{FP} are positive definite. Thus in what follows we shall assume (16).

Langevin Approach

It is well known in statistical mechanics that the system described by Fokker-Planck equation is also described equivalently in terms of Langevin equation.[4] The equivalence between these two approaches can be proven in various ways,[5] but we present a simple heuristic method following reference 6 which is useful to derive the Langevin equation for other theories such as lattice gauge theories.

In the Langevin approach, one considers a field in d + 1 dimensional space and set up a Langevin equation, which is a first order differential equation with respect to the fictitious time τ and contains a random source term. The statistical average is then defined by the average over the random source function. We shall use $\eta_\ell(x,\tau)$ for the random source function and call the average over η as the η-average:

$$\langle \ldots \rangle_\eta \equiv \lim_{\Lambda\to\infty} \frac{\int D\eta \ldots e^{-\frac{1}{4}\int dx d\tau d\tau' a_\Lambda(\tau-\tau')\eta_\ell(x,\tau)\eta_\ell(x,\tau')}}{\int D\eta \; e^{-\frac{1}{4}\int dx d\tau d\tau' a_\Lambda(\tau-\tau')\eta_\ell(x,\tau)\eta_\ell(x,\tau')}} \tag{17}$$

where $a_\Lambda(\tau-\tau')$ is a regulator function[7] such that

$$\int a_\Lambda(\tau-\tau') \, d\tau' = 1$$
$$\lim_{\Lambda\to\infty} a_\Lambda(\tau-\tau') = \delta(\tau-\tau') \tag{18}$$

We note also a_Λ is a symmetric function:

$$a_\Lambda(\tau) = a_\Lambda(-\tau) \tag{19}$$

Let $\phi_\ell^\eta(x,\tau)$ be a solution of Langevin equation. The connection between the Fokker-Planck and Langevin approach is then given by

$$\int D\,\phi\; F[\phi(\cdot)]\; P[\phi,\tau] = \langle F[\phi^\eta(\cdot,\tau)] \rangle_\eta \qquad (20)$$

where $F[\phi(\cdot)]$ is an arbitrary functional of $\phi_\ell(x)$. Based on this identification we derive the Langevin equation which corresponds to the previously given Fokker-Planck Hamiltonian (12) and (16).

<u>Derivation of Langevin Equation</u>

Since the suffix ℓ is superfluous, we omit it, or equivalently we write x for x and ℓ. We also omit $[\phi(\cdot)]$ for functional. We simply write F for $F[\phi(\cdot)]$ and $F(\tau)$ for $F[\phi(\cdot,\tau)]$ etc.

Taking a time derivative of (20) and using the Fokker-Planck equation we obtain

$$\begin{aligned} -\frac{\partial}{\partial\tau}\int D\,\phi\; F\; P(\tau) = \int D\,\phi\; P(\tau)\int dxdy\,\Big\{\frac{\delta S}{\delta\phi(x)}\,K(x,y)\,\frac{\delta F}{\delta\phi(y)} \; - \\ - \frac{\delta}{\delta\phi(x)}\,K(x,y)\,\frac{\delta F}{\delta\phi(y)}\Big\} \end{aligned} \qquad (21)$$

By using (16) the second term of (21) can be written as

$$\begin{aligned} -\int D\,\phi\; P(\tau)\int dxdydz'\,\Big\{\frac{\delta G(x,z)}{\delta\phi(x)}\,G(y,z)\,\frac{\delta F}{\delta\phi(y)} \; + \\ + \; G(x,z)\,\frac{\delta}{\delta\phi(x)}\Big(G(y,z)\,\frac{\delta F}{\delta\phi(y)}\Big)\Big\} \end{aligned}$$

Therefore,

$$- \frac{\partial}{\partial \tau} \int D\,\phi\,F\,P(\tau)$$

$$= \int D\,\phi\,P(\tau)' \left\{ \int dx\, A(x) \frac{\delta F}{\delta \phi(x)} - \int dxdydz\, G(x,z) \frac{\delta}{\delta \phi(x)} \left(G(y,z) \frac{\delta F}{\delta \phi(y)}\right) \right. \tag{22}$$

where A(x) is given by

$$A(x) = \int dy\, K(x,y) \frac{\delta S}{\delta \phi(y)} - \int dz\, G(x,z) \frac{\delta G(y,z)}{\delta \phi(y)} \tag{23}$$

Using (20) we can write equation (22) in terms of ϕ^η and the η-average as

$$- \frac{\partial}{\partial \tau} < F >_\eta$$

$$= < \int dx\, A(x) \frac{\delta F}{\delta \phi^\eta(x,\tau)} - \int dxdydz\, G(y,z) \frac{\delta}{\delta \phi^\eta(y,\tau)} \left(G(x,z) \frac{\delta F}{\delta \phi^\eta(x,y)}\right) > \tag{24}$$

The left hand side of (24) is given by

$$< \int dx \left(- \frac{\partial \phi^\eta(x,\tau)}{\partial \tau}\right) \frac{\delta F}{\delta \phi^\eta(x,\tau)} >_\eta$$

so that one set

$$- \frac{\partial \phi^\eta(x,\tau)}{\partial \tau} = A(x;\, \phi^\eta(\cdot,\tau)) - \int dz\, G(x,z;\, \phi^\eta(\cdot,\tau))\, \eta(z,\tau) \tag{25}$$

since if $\phi^\eta(x,\tau)$ is a retarded solution of (25), i.e.

$$\frac{\delta \phi^\eta(x,\tau)}{\delta \eta(y,\, \tau')} = 0 \quad \text{for} \quad \tau' > \tau \tag{26}$$

then

$$< \eta(x,\tau)\ M\ [\phi^{\eta}(\cdot,\tau)] >_{\eta} \ = \int dy \ < G(y,x)\ \frac{\delta M}{\delta\phi^{\eta}(y,\tau)} >_{\eta} \qquad (27)$$

Equation (25) is called the Langevin equaiton.

Proof of (27):

$$< \eta(x,\tau)\ M >_{\eta} \ = \lim_{\Lambda\to\infty} \int d\tau' 2\ a_{\Lambda}^{-1}\ (\tau-\tau')\ < \frac{\delta}{\delta\eta(\tau-\tau')}\ M >_{\eta}$$

$$= \lim_{\Lambda\to\infty} \int d\tau'\ 2\ a_{\Lambda}^{-1}(\tau-\tau')\ < \int dy\ \frac{\delta\phi^{\eta}(y,\ \tau)}{\delta\eta\ (x,\ \tau')}\ \frac{\delta M}{\delta\phi^{\eta}(y,\tau)} >_{\eta}$$

where a_{Λ}^{-1} is an inverse function of a_{Λ}:

$$\int a_{\Lambda}^{-1}(\tau-\tau')\ a_{\Lambda}(\tau'-\tau'')\ d\tau' = \delta(\tau-\ \tau'').$$

Next we prove

$$\lim_{\Lambda\to\infty} \int\ a_{\Lambda}^{-1}\ (\tau-\tau')\ d\tau'\ \frac{\delta\phi^{\eta}(y,\tau)}{\delta\eta(x,\tau')} \ = \frac{1}{2}\ G(y,x) \qquad (28)$$

then (27) follows. Integrating the Langevin equation we obtain

$$\phi^{\eta}(x,\tau) = \phi_0(x,0) - \int_0^{\tau}\ A(x,\phi^{\eta}(\cdot,\ \tau''))\ d\ \tau''$$

$$+ \int_0^{\tau} d\ \tau'' \int dy\ G(x,y;\ \phi^{\eta}(\cdot\tau''))\ \eta(y,\ \tau'')$$

Thus,

$$\frac{\delta\phi^{\eta}(x,\tau)}{\delta\eta(y,\tau')} = G(x,y)\ \theta(\tau-\tau') + \int_0^{\tau} d\tau''\ \frac{\delta\phi^{\eta}(z,\tau'')}{\delta\eta(y,\tau')}\ (\ \cdots\ )$$

Inserting it into (28) and using

$$\lim_{\Lambda\to\infty} \int_0^{\tau} a_{\Lambda}^{-1}(\tau-\tau')\ \theta(\tau-\tau')\ d\tau' = \frac{1}{2}$$

$$\lim_{\Lambda\to\infty} \int_0^{\tau} d\tau' \int_0^{\tau} d\tau''\ a_{\Lambda}^{-1}\ (\tau-\tau')\ \theta\ (\tau''-\tau') = 0$$

we obtain (28).

Parisi-Wu Stochastic Quantization

One solves Langevin equation

$$\begin{aligned}\frac{\partial\phi_{\ell}(x,\tau)}{\partial\tau} = &- \int dx' \sum_{\ell'} K_{\ell\ell'}\ (x,x';\ \phi(\cdot,\tau))\frac{\delta S[\phi(\cdot,\tau)]}{\delta\phi_{\ell'}(x',\tau)} \\ &+ \sum_{m} \int dx'\ G_{\ell m}\ (x,x';\ \phi(\cdot,\tau)) \int dx'' \sum_{\ell'} \frac{\delta G_{\ell' m}(x',x'';\ \phi(\cdot,\tau)}{\delta\phi_{\ell'}(x')} \\ &+ \sum_{m}\int dx'\ G_{\ell m}\ (x,x';\phi(\cdot,\tau))\ \eta_m(x',\tau)\end{aligned} \tag{29}$$

with retarded condition (26). Then one takes the η-average of the product of ϕ^n and take $\tau \to \infty$ limit to obtain Euclidean Green's function.

As we see from our derivation of Langevin equation, $G_{\ell m}(x,y;\phi)$ is entirely arbitrary. We can choose any form depending on the conveniences. The simplest choice $G_{\ell m}(x,y;\phi)=\delta_{\ell m}\ \delta(x-y)$ leads the standard Langevin equation given in reference 1.

Langevin Equation for Lattice Gauge Theories[8]

Since Wilson's lattice gauge theory consists of the dynamical link variables which are U(N) matrix, it is sufficient to derive Langevin equation for U(N) matrix. This can be done by modifying the previous discussions as follows:

$$\begin{array}{lll} \text{integration measure} & \Rightarrow & \text{Haar measure} \\ D\phi & \to & dU \\ \text{derivative} & \Rightarrow & \text{left Lie derivative} \\ -i\,\dfrac{\delta}{\delta\phi(x)} & \to & E_\alpha \end{array}$$

such that

$$E_\alpha(U) = (t_\alpha\ U) \tag{30}$$

Explicit Construction of E_α: Let t_α ($\alpha = 1,2,\ldots, N^2$) be N x N fundamental representation of U(N) Lie algebra:

$$\begin{aligned} [t_\alpha, t_\beta] &= i\ C^\gamma_{\alpha\beta}\ t_\gamma \\ (t_\alpha)_{ij}\ (t_\alpha)_{k\ell} &= \delta_{i\ell}\ \delta_{jk} \end{aligned} \tag{31}$$

Let us parametrize N x N unitary matrix U by

$$U(A) = \exp i\ t_\alpha\ A_\alpha \equiv e^{i\underset{\sim}{A}} \tag{32}$$

$$\frac{\partial U(A)}{\partial A_\alpha} = i \int_0^1 da\, e^{ia\underset{\sim}{A}} t_\alpha\, e^{i(1-a)\underset{\sim}{A}}$$

$$= i \int_0^1 da\, e^{ia\underset{\sim}{A}} t_\alpha\, e^{-ia\underset{\sim}{A}} U(A)$$

$$= i \int_0^1 da\, \mathrm{tr}\, (e^{ia\underset{\sim}{A}} t_\alpha\, e^{-ia\underset{\sim}{A}} t_\beta)\, t_\beta\, U(A)$$

$$= i\, N_{\alpha\beta}(A)\, t_\beta\, U(A) \tag{33}$$

where

$$N_{\alpha\beta}(A) = \int_0^1 da\, \mathrm{tr}\, (e^{ia\underset{\sim}{A}} t_\alpha\, e^{-ia\underset{\sim}{A}} t_\beta) \tag{34}$$

Let

$$E_\alpha = i\, N^{-1}_{\alpha\beta} \frac{\partial}{\partial A_\beta} \tag{35}$$

then

$$E_\alpha = (U_{ij}) = (t_\alpha U)_{ij} \tag{30}$$

One can confirm E_α satisfies the U(N) commutation relations:

$$[E_\alpha, E_\beta] = i\, C^\gamma_{\alpha\beta}\, E_\gamma \tag{36}$$

Fokker Planck Hamiltonian:

$$H_{FP} = \sum_{x,\mu} E_{\alpha,\mu}(x) \left(E_{\alpha,\mu}(x) + \left(E_{\alpha,\mu}(x)\, S\right)\right) \tag{37}$$

The first factor $E_{\alpha,\mu}(x)$ is for the probability conservation while the second factor for

$$P(\tau) \underset{\tau\to\infty}{\to} e^{-S} \Big/ \int dU\, e^{-S}$$

Derivation of Langevin Equation: Assume that U depends on a parameter τ. Note the identity:

$$\frac{\partial}{\partial\tau} F[U] = tr\left(\frac{\partial U}{\partial\tau} U^{-1} \underset{\sim}{E}\right) F[U] \tag{38}$$

where

$$\underset{\sim}{E} = t_\alpha\, E_\alpha \tag{39}$$

Using the same method used for Bose fields we obtain the following Langevin equation

$$-i \left(\frac{\partial}{\partial\tau} U \right) U^{+} = -i\, \underset{\sim}{E}(S[U]) = \underset{\sim}{\eta} \tag{40}$$

where

$$\underset{\sim}{\eta} = t_\alpha\, \eta_\alpha$$

$$< \eta_\alpha\, \eta_\beta >_\eta = 2\, \delta_{\alpha\beta} \tag{41}$$

The Langevin equation (40) can be used for any U(N) matrix model. Indeed, it has been used by Guha and Lee[9] for the numerical computation of free energy of chiral SU(N) x SU(N) in 2,3, and 4 dimensions in order to compare the stochastic quantization method with the other methods such as Monte Carlo.

Stochastic Quantization of Fermi Fields

The stochastic quantization methods for Fermi fields so far discussed in the literatures are classified into two: the one[10] is to use the boson field method discussed above after integrating all the Fermi fields out, and the other[11] is to derive Langevin equation for anticommuting Fermi fields. Although the first method is more appealing because of its possible applications to numerical calculations, we shall confine our discussion to the latter method. The essential part of the method we present here is due to reference 12 .

The path integral to be considered is

$$\int D\psi\, D\bar{\psi}\ (\cdots\cdot)\ e^{-S[\psi,\bar{\psi}]} \Big/ \int D\psi\ D\bar{\psi}\ e^{-S[\psi,\bar{\psi}]} \tag{37}$$

We use ψ and $\bar{\psi}$ for anticommuting variables. We restrict S to a bilinear form

$$S = \int \bar{\psi}\,(x)\ G\ \psi\,(x)\ d\,x \tag{38}$$

since we can always write S in this form by using auxiliary scalar fields. Thus G may contain scalar fields as well as derivative operators.

An appropriate form of Fokker-Planck Hamiltonian is given by

$$H_{FP} = \int dx \, [\frac{\delta}{\delta\psi(x)} G^{\dagger}(\frac{\delta}{\delta\bar{\psi}(x)} + \frac{\delta S}{\delta\bar{\psi}(x)}) - \frac{\delta}{\delta\bar{\psi}} G (\frac{\delta}{\delta\psi} + \frac{\delta S}{\delta\psi(x)})] \, .$$

$$= \int dx \, [\frac{\delta}{\delta\psi(x)} G^{\dagger}(\frac{\delta}{\delta\bar{\psi}(x)} + G \, \psi(x)) - \frac{\delta}{\delta\bar{\psi}} G (\frac{\delta}{\delta\psi} - G^{\dagger} \, \bar{\psi}(x))] \tag{39}$$

It is possible to show that the eigenvalues of H_{FP} are positive definite.

A proof of positivity goes as follows. We first note that as far as integration (37) is concerned $\bar{\psi}$ is an independent variable of ψ. One can disregard the origin of $\bar{\psi}$ and introduces two independent Fermi operators $\hat{\psi}$ and $\hat{\bar{\psi}}$ and its conjugates $\hat{\psi}^{\dagger}$ and $\hat{\bar{\psi}}^{\dagger}$:

$$\{ \hat{\psi} \, , \, \hat{\psi}^{\dagger} \} = 1 \quad , \quad \{ \hat{\bar{\psi}} \, , \, \hat{\bar{\psi}}^{\dagger} \} = 1$$
$$\{ \hat{\psi} \, , \, \hat{\bar{\psi}}^{\dagger} \} = 0 \quad \text{etc.} \tag{40}$$

Denoting the standard number representation by $|i,\bar{j}>$ $(i,j = 0$ or $1)$, we construct a coherent representation by

$$|\psi,\bar{\psi}> = |0,\bar{0}> + \psi \, |1,\bar{0}> + \bar{\psi} \, |0,\bar{1}> + \psi\bar{\psi} \, |1,1>$$

where ψ and $\bar{\psi}$ are the anticommuting variables used in the integration. Note

$$\begin{aligned} \hat{\psi}^{\dagger} \, |\psi,\bar{\psi}> &= \frac{\delta}{\delta\psi} \, |\psi,\bar{\psi}> \\ \hat{\psi} \, |\psi,\bar{\psi}> &= \psi \, |\psi,\bar{\psi}> \\ \hat{\bar{\psi}}^{\dagger} \, |\psi,\bar{\psi}> &= \frac{\delta}{\delta\bar{\psi}} \, |\psi,\bar{\psi}> \\ \hat{\bar{\psi}} \, |\psi,\bar{\psi}> &= \bar{\psi} \, |\psi,\bar{\psi}> \end{aligned} \tag{41}$$

So the Fokker-Planck Hamiltonian (39) is the coherent representation of

$$H_{FP} = \int dx\, [\hat{\psi}^{\dagger} G^{\dagger}(\hat{\bar{\psi}}^{\dagger} + G\hat{\psi}) - \hat{\bar{\psi}}^{\dagger} G(\hat{\psi}^{\dagger} - G^{\dagger}\hat{\bar{\psi}})] \tag{42}$$

By making a similarity transformation defined by

$$\hat{H}_{FP} = e^{-\int dx\, \hat{\bar{\psi}}^{\dagger} G^{-1} \hat{\psi}^{\dagger}}\, H_{FP}\, e^{\int dx\, \hat{\bar{\psi}}^{\dagger} G^{-1} \hat{\psi}^{\dagger}} \tag{43}$$

we obtain

$$\hat{H}_{FP} = \int dx\, (\hat{\bar{\psi}}^{\dagger} G^{\dagger} G \hat{\psi} + \hat{\psi}^{\dagger} G G^{\dagger} \hat{\bar{\psi}}) \tag{44}$$

which has a positive definite form.

We use the entirely same method used in the previous sections in order to derive the following Langevin equation:

$$\begin{aligned} \frac{\partial}{\partial\tau} \psi(x,\tau) &= -G^{\dagger}G\ \psi(x,\tau) + G\,\eta\,(x,\tau) \\ \frac{\partial}{\partial\tau} \bar{\psi}(x,\tau) &= GG^{\dagger}\ \bar{\psi}(x,\tau) + \bar{\eta}\,(x,\tau) \end{aligned} \tag{45}$$

Note $\bar{\psi}$ is an independent variable and two equations of (45) are independent also. η and $\bar{\eta}$ are a pair of anticommuting noise functions such that

$$\begin{aligned} < \eta_{\alpha}(x,\tau)\, \bar{\eta}_{\beta}(x',\tau') >_{\eta} &= - < \bar{\eta}_{\beta}(x',\tau')\, \eta_{\alpha}(x,\tau) >_{\eta} \\ &= 2\delta_{\alpha\beta}\, \delta(x-x')\, \delta(\tau-\tau') \end{aligned} \tag{46}$$

It has been shown by K. Ishikawa[13] that a supersymmetric gauge theory can also be formulated in a similar way.

ochastic Regularization

It has been proposed by Breit, Gupta and Zaks[7] that the theory is gularized by using the stochastic noise regulator a_Λ. Since this regulates ly the fictitious time direction, at first sight it does not affect the variances of the theory such as gauge and Lorentz invariance. However, r the finite Λ the Langevin process is no longer Markovian so that it is t evident that all the nice properties obtained by using Fokker-Planck uation should hold.

knowledgements

I acknowledge the valuable discussions with J. Alfaro and K. Ishikawa.[13] st of the ideas are derived from the discussions with them.

ferences

J. Alfaro and B. Sakita, p.65, Gauge Theory and Gravitation, Proc. Nara, Japan, ed. K. Kikkawa et al, Springer-Verlag (1983).

J. Alfaro and B. Sakita, Proc. Topical Symposium on High Energy Physics, ed. T. Eguchi and Y. Yamaguchi,World Scientific 1983.

G. Parisi and Y. Wu, Sci. Sin. 24, 483 (1981).

e.g. Nelson Wax, Selected Papers on Noise and Stochastic Processes, Dover Publications, New York, NY (1954)

R.F. Fox, Phys. Reports 48, 179-283 (1978) and references therein.

E. Gozzi, CCNY-HEP-83/4 and references therein.

F. Langouche, F. Roekaerts and E. Tirapegui, Prog. Theor. Phys. 61, 161 (1979).

J. Alfaro and B. Sakita, ref.2

J.D. Breit, S. Gupta and A. Zaks, ISA Preprint, 1983.

8. I.T. Drummond, S. Duane and R.R. Horgan, Nucl Phys. B220, 119 (1983).

J. Alfaro and B. Sakita, ref.2.

A. Guha and S.-C. Lee, Phys. Rev. D27, 2412 (1983).

M.B. Halpern, UCB-PTH-83/1

J. Alfaro, Ph.D. Thesis (1983).

9. A. Guha and S.-C. Lee (private communication)

10. J.R. Klauder, Bell Lab. Preprint, 1983.

D. Zwanziger, Phys. Rev. Lett. 50, 1886 (1983).

11. Y. Kakudo, Y. Taguchi, A. Tanaka and K. Yamamoto, OS-GE-82-39.

T. Fukui, H. Nakazato, I. Ohba, K. Okano and Y. Yamanaka, WU-HEP-82-7.

J.D. Breit, S. Gupta and A. Zaks, ref.7.

12. T. Fukui et al. ref.11

13. K. Ishikawa (in preparation)

STOCHASTIC PROCESSES IN LATTICE SUPERSYMMETRY

Sergio Cecotti
Scuola Normale Superiore, Pisa

ABSTRACT

We study the conditions under which a supersymmetric theory is equivalent (at the quantum level) to a classical stochastic system (specified by a set of Langevin equations), both on the continuum and on the lattice. In particular we discuss how N=2 two dimensional models can be put on the lattice preserving the stochastic equivalence.

CONTENTS

1. INTRODUCTION: LATTICE SUPERSYMMETRY.

As is well known[1)], the supersymmetry algebra, in the continuum case, is just the graded version of the Poincarè group.

$$\{Q_\alpha, \bar{Q}_\beta\} = 2 \not{P}_{\alpha\beta} . \tag{1.1}$$

(*) ERS = Elitzur, Rabinovici and Schwimmer, see ref. (2).

Obviously, on a lattice we have only a discrete subgroup of Poincarè transformations. Given the mathematical fact that a discrete group cannot be graded, on a (space-time) lattice it is not possible to maintain the standard relation between fermionic symmetries and translations.

Then, if we want to keep -on the lattice- a direct contact between supersymmetry and the (lattice) translations, these have to contain a continuous subgroup. The minimal choice -just one continuous translation- leads us, naturally, to consider the case of a Hamiltonian lattice, i.e. continuous time and discrete space. In this approach to lattice supersymmetry, the Hamiltonian is the only bosonic generator of the superalgebra. Moreover, this kind of lattice is particularly suited for our purposes, given that the discussion of the process in terms of Langevin equations (which, by the way, are differential equations in time) is Hamiltonian in character.

Thus, we have to look for a lattice superalgebra containing only H and fermionic operators.

The obvious choice for a lattice supersymmetry Algebra is just one supercharge Q, linear in the canonical fermionic operators, such that

$$Q^2 = H \tag{1.2}$$

which will give some interesting models in 2 dim.[2)].

From the Hamiltonian point of view, there is a perhaps more natural superalgebra; two supercharges linear in the fermionic operators, Q_A, A = 1, 2

$$\{Q_A, Q_B\} = 2\,\delta_{AB} H, \tag{1.3}$$

which is the algebra of supersymmetric Quantum Mechanics[3], related to the superfield approach, in the sense that the corresponding models are deducible from a superpotential (see later). Note that in the minimal case, 1 bosonic + 1 fermionic degrees of freedom, eq. (1.2) implies eq. (1.3). Indeed, eq. (1.3) is satisfied with $Q_1 = Q$ and $Q_2 = i(-)^F Q$ (*). In the non-minimal case, this Q_2 is <u>not</u> linear.

Eq. (1.3) will play the central role in our discussion.

Another valuable virtue of a Hamiltonian lattice is the existence of a Witten index[4]

$$\Delta = \mathrm{Str}\exp(-\beta H) = n_B^0 - n_F^0 .$$

This fact can hardly be over-emphasized. It implies equal numbers of fermionic and bosonic (non-zero-energy) states and <u>identical</u> dispersion relations. This fact has non-trivial consequences for the fermion doubling problem[5]. If the bosonic kinetic term has the standard form -i.e. no "bosonic" doubling-, lattice supersymmetry tells us that there is no fermionic doubling as well, and hence supersymmetry must imply some chiral breaking terms in the Lagrangian -such as, for instance, Wilson's terms[6]- in agreement with the general argument[5]. So, lattice supersymmetry gives interesting insights on both lattice fermions and continuum supersymmetry, as we shall see.

2. LOCAL NICOLAI MAPPINGS AND THE STATIONARY HJ EQUATION.

In this talk, we are interested in a special subclass of lattice supersymmetry models: those which are equivalent, in the sense of quantum theory, to a stochastic process governed by a Langevin-like equation.

It can be shown[7] -perturbatively- that a supersymmetric theory

---

(*) $(-)^F$ is the grading operator of the superalgebra.

(with $\Delta \neq 0$) the quantum functional measure

$$[d\Phi]\ \exp -S(\Phi)$$

can be transformed into a Gaussian measure -over bosons only- with covariance 1, with a suitable non-local change of field variables. In a few cases, this can be checked non-perturbatively.

In certain specific cases this change of variables (called Nicolai mapping) is local in space and time. In these cases, we have some fields $h_i(\varphi_j)$ such that

$$\tfrac{1}{2}\Sigma_i h_i^2 = L_B + \text{surface terms} \tag{2.1}$$

where L_B is the bosonic part of the Euclidean Lagrangian

$$L_B = L|_{\psi = 0}.$$

Moreover, the Jacobian of the transformation is

$$\det[\delta h_i/\delta\varphi_j] = \int [d\psi_i d\bar{\psi}_j] \exp -\int \bar{\psi} D_F[\varphi]\psi, \tag{2.2}$$

where $L = L_B + \bar{\psi}\, D_F\, \psi.$

From eqs. (2.1, 2) we have

$$\int [d\varphi d\psi d\bar{\psi}]\, e^{-S} = \int [dh_i]\, e^{-\frac{1}{2}\int \Sigma_i h_i^2 dx} \tag{2.3}$$

if we assume periodic boudary conditions[8).]

Examples of this situation in 2 dim. were found by Parisi and

Sourlas[9] and, independently, by the authors of ref. 10.

The general Ansatz for the local mapping with a (Euclidean) Lagrangian of the form

$$L = \frac{1}{2}\sum_i \dot{q}_i^2 + U(q_i) + \bar{\psi} D_F(q) \psi \tag{2.4}$$

is very simple. h_i must be linear in $\dot{q}_j$,

$$h_i = A_{ij}\dot{q}_j + F_i(q).$$

From eqs. (2.1,4) it follows that A_{ij} is orthogonal. Redefining $h'_i = A^{-1}_{ij}h_j$, we get

$$h_i = \dot{q}_i + G_i(q). \tag{2.5}$$

Requiring that L has no term linear in $\dot{q}_i$, we get

$$F_i(q) = \frac{\partial W}{\partial q_i} = \partial_i W. \tag{2.6}$$

Then,

$$\frac{1}{2}\sum_i h_i^2 = \frac{1}{2}\sum_i \dot{q}_i^2 + \frac{1}{2}\sum_i (\partial_i W)^2 + \frac{dW}{dt} . \tag{2.7}$$

Comparing eq. (2.1,4,7) we obtain a differential equation for W

$$\frac{1}{2}\sum_i (\partial_i W)^2 - U(q_i) \equiv E = 0, \tag{2.8}$$

which can be thought as the stationary (i.e. with $\partial W/\partial t = 0$) Hamilton-Jacobi (HJ) equation for the Lagrangian L_B. As is well known, this equation describes the zero-(Euclidean)-energy manifold[10]. If we consider eq. (2.5) as a Langevin stochastic equation, eq. (2.8) becomes very suggestive. Indeed, it amounts to the usual relation[11) between a particular solution of the HJ equation and the drift coefficients of the Langevin process.

The fact that we are looking for a stationary solution is rather natural, if we recall that our process has to describe the VACUUM of the quantum theory, i.e. a time-translational invariant state. For this reason, the solution of eq. (2.8), having the property of generating a Nicolai mapping, was called in ref. 10 the "classical process of the Euclidean vacuum".

We stress that the equivalence with a zero-energy stochastic process is another explanation of the non-renormalization of the vacuum energy in an unbroken supersymmetric theory.

At this point, the main problem is under what circumstances eq. (2.8) does admit solutions W which are local (in space) functionals of the fields, as needed in order to have a fully local mapping. It is not a surprise that these "local" integrability conditions are strictly connected with the doubling problem.

Suppose that we want to describe, on a 2 dim. lattice, the model,

$$\mathcal{L} = \frac{1}{2}(\partial_\mu \varphi)^2 + \bar{\psi}_L \not\partial \psi_L \qquad (\varphi \text{ real}) \; . \tag{2.9}$$

In this case, eq. (2.8) becomes

$$\int \left\{ \left(\frac{\delta W}{\delta \varphi} \right)^2 - (\partial_1 \varphi)^2 \right\} = 0 \tag{2.10}$$

whose local solutions have the property

$$\frac{\delta W}{\delta \varphi} = \pm\, \partial_1 \varphi \tag{2.11}$$

which cannot be integrated, because it would imply that

$$\frac{\delta^2 W}{\delta\varphi(x)\,\delta\varphi(y)} = \partial_1 \delta(x-y), \tag{2.12}$$

which is absurd, because the l.h.s. of eq. (2.12) is symmetric for $x \leftrightarrow y$, whereas the r.h.s. is antisymmetric; (this argument holds, mutatis mutandis both in the continuum and on the lattice).

Indeed, if a local solution to eq. (2.10) would exist, we would have a paradox, because one could introduce a neutrino, ψ_L on the lattice, working in exactly the same way as we shall do for the models discussed in this talk, contrary to the general theorem of ref. 5.

Of course, we could formulate the relevant integrability conditions directly in terms of the functional-differential structure of the HJ equation, but we think that it is better to give an algebraic interpretation of these conditions.

3. THE ALGEBRAIC MEANING OF THE LOCAL INTEGRABILITY CONDITIONS(*).

In this section, we prove that the quantity W in eq. (2.6) is just what is called the "superpotential" of 1D supersymmetry. THIS SUPERPOTENTIAL IS LOCAL IF THE SUPERCHARGES ARE LOCAL (in the sense that they are the space integrals of local currents).

Of course, whether a theory admits or not a superpotential, depends only on the algebra. Apart from certain subtleties which play no role in the case we are interested in, we have the following

(*) See ref. 12

THEOREM[12,13]: for each system with the superalgebra in eq. (1.3) one can reconstruct a superpotential functional W(**). W is local iff Q_A is.

Here comes the argument. Suppose we have a theory with K bosonic and K fermionic degrees of freedom

$$[x_i, p_j] = i\delta_{ij}, \quad \{\chi^i_A, \chi^j_B\} = 2\delta_{AB}\delta^{ij} \tag{3.1}$$

having two Hermitian operators Q_A, linear in the canonical fermionic operators χ, so that the algebra (1.3) holds with a Hamiltonian

$$H = \frac{1}{2}\sum_i p_i^2 + U(x_i, \chi^i_A)\,. \tag{3.2}$$

We can show[12,13] that, without loss of generality—apart from considerations which do not apply to canonical ungauged models in 2 or 4 dim.—, the general Ansatz for Q_A is

$$\sqrt{2}\, Q_A = \chi^i_A p_i + \chi^i_A R^A_i(x) + \epsilon_{AB}\chi^i_B S^{(A)}_i(x) \tag{3.3}$$

and then,

i) Q_A^2 does not contain terms linear in p_i, requires $R_i^{(A)}=0$

ii) $Q_1^2 = Q_2^2$ implies $S_I^{(A)}(x) = S_i(x)$;

iii) $0 = \{Q_1, Q_2\} = i/2(\chi^i_1\chi^j_1 - \chi^i_2\chi^j_2)(\partial_i S_j - \partial_j S_i)$ implies $S_j(x) = \partial_j W(x)$.

This completes the proof of the existence of W if the algebra (1.3) holds.

At this point, we can compute the Hamiltonian H and the

(**) A more general analysis can be found in ref. 14.

(Euclidean) Lagrangian L_E from eq. (1.3)

$$H = \frac{1}{2}\sum_i [p_i^2 + (\partial_i W)^2] + \frac{i}{2}\sum_{ij} \partial_i\partial_j W \chi_1^i \chi_2^j \tag{3.4}$$

$$L_E = \frac{1}{2}\sum_i [\dot{x}_i^2 + (\partial_i W)^2] + \sum_{ij} \bar{\xi}_i [\delta_{ij} d_t + \partial_i\partial_j W] \xi_j \tag{3.5}$$

where $\bar{\xi}_i$ and ξ_j are Grassman fields.

Now, it is obvious from eq. (3.5) that the functional mapping

$$h_i = \dot{x}_i + \partial_i W \tag{3.6}$$

works. Indeed, comparing with eq. (3.5), we have

$$\frac{1}{2}\sum_i h_i^2 = L_B + \frac{dW}{dt} \tag{3.7a}$$

and

$$\begin{aligned} \det[\delta h_i / \delta x_j]_{per} &= \det[\delta_{ij} d_t + \partial_i \partial_j W]_{per} = \\ &= \int [d\bar{\xi} d\xi]_{per} \exp\left(-\int_0^\beta d\tau L_F(\bar{\xi}\xi, x)\right). \end{aligned} \tag{3.7b}$$

Note that the superpotential i̲s̲ a solution of the stationary Hamilton-Jacobi equation, eq. (2.8).

Now we sketch how this integrability condition—in its algebraic form, eq. (1.3)—can be used in order to construct models, HAVING A STOCHASTIC INTERPRETATION, in dimension larger than one, in the continuum case. Explicit examples are presented in refs. 9 and 10. These remarks will be used to understand better the relationships between continuum and lattice models and, moreover, they show explicitly that local charges imply local W[12].

The idea is very simple. From the considerations of refs. 8 and 10 we know that we must put the theory in a Euclidean box with periodic boundary conditions. Then, —in the continuum— we decompose the superfields into Fourier components (in space only). With a suitable UV regularization (see refs. 10, 12 and 13), we have a situation with a finite number of degrees of freedom, as in the Quantum Mechanical case, BUT with a major difference with respect to the previous situation: the superalgebra is <u>not</u> eq. (1.3) but eq. (1.1) and, so the argument for the existence of a local W does not work.

Our strategy is then the following[12,13]. First, Algebra (1.1) can be brought into the form of eq. (1.3); this <u>is</u> possible if N=2. Then, using the arguments of this section, we can reconstruct an effective superpotential W, which, when substituted in eq. (3.6), gives us the Nicolai mapping for the regularized version of the theory.

Note that the drift coefficients in eq. (3.6) —$\partial_i W$— are just the coefficients of $\varepsilon_{AB}\chi^i{}_B$ in Q_A, as follows from eq. (3.3). Then, if Q_A is a local functional in the field, so is the mapping (3.6).

(The mapping has no dependence on the cut-off, so we can remove it without problems, recovering the mapping for the unregularized version of the theory.)

The reduction from the algebra of N=2, 2 dim., supersymmetry

$$\{Q^a_\alpha, Q^b_\beta\} = 2\,\delta^{ab}(\delta_{\alpha\beta} H - \sigma^3_{\alpha\beta} P) \tag{3.8}$$

(Majorana Basis; a,b,α,β = 1,2)

to the algebra in eq. (1.3) can be done as follows. Consider

$$\tilde{Q}_A = 2^{-1/2}(Q_1{}^A + Q_2{}^A).$$

Then, from eq. (3.8), we get

$$\{\tilde{Q}_A, \tilde{Q}_B\} = 2\delta_{AB} H \tag{3.9}$$

satisfying eq. (1.3) (also the other technical assumptions are fulfilled, see ref. 13).

Hence, D=2 N=2 supersymmetric models, of the class having linear supercharges (i.e. no 4-fermion interactions) do have a mapping, The corresponding W is[10)]

$$W[\varphi_i, \varphi_i^*] = \\ = \int dx \left[\frac{i}{2} \sum_i \varphi_i^* \overleftrightarrow{\partial}_1 \varphi_i + 2 \,\mathrm{Re}\, V(\varphi_i) \right], \tag{3.10}$$

where V (the superpotential in the standard sense) is an analytic function of ϕ_i.

Interestingly enough, this class of models is completely UV finite. Indeed, they belong to a class analyzed in detail in ref. 15. The same result will be obtained by "infradiagrammatic"[10)] analysis.

The converse is also true; if a local mapping exists, there is a subalgebra isomorphic to that of eq. (1.3); because this is not a subalgebra of N=1 SUSY, for a (2 dim.) Lorentz invariant model we need N=2.

The same is true for D=4. In this case,

$$\{ Q^a_\alpha, Q^b_\beta \} = 2\delta^{ab} (\delta_{\alpha\beta} H - \vec{\alpha}_{\alpha\beta} \vec{P}). \tag{3.11}$$

(a,b = 1,2; α,β = 1,..,4)

In the Majorana basis we use[16)], $\tilde{Q}_A = 2^{-1/2}(Q^1{}_A + Q^2{}_{A+2})$ does fulfill the algebra (3.9). Hence, if we have N=2 without central charges there is a local mapping even in 4 dim. Unfortunately, there is not a single, non-free, consistent model in this class -ungauged N=2 SUSY without central charges- in 4 dim. We are left with only the free massless case.

We close this section with the following remark, for future reference. If W is a (local) solution of eq. (2.8) so it is -W. From eq. (3.6), we see that the interchange of W with -W corresponds to changing the sign of time.

$$h_i = \dot{x}_i - \partial_i W$$

is also a mapping, provided we make the change in eq. (3.5) $\bar{\xi}_i \leftrightarrow \xi_i$ (see ref. 10). Conventionally, we will refer to this as the "backward process".

4. THE ERS[(*)] N=1 MODELS.

First of all, we review the models of ref. 2, which have the supersymmetry algebra of eq. (1.2) and have no mapping.

In the continuum case there are two supercharges with the property (1.2). We consider the linear combinations $Q_\pm$ such that

$$Q_\pm^2 = H \qquad \{Q_+, Q_-\} = P$$

their explicit expression is

$$Q_\pm = \int dx \{\pi(x)\psi_{1,2}(x) + \partial_1\varphi(x)\psi_{2,1}(x) \pm V(\varphi(x))\psi_{2,1}(x)\}. \tag{4.1}$$

Now discretize in a straightforward way one of the two, say Q_+

$$Q = \sum_x \{\pi(x)\psi_1(x) + \Delta^+\varphi(x)\psi_2(x) + V(\varphi(x))\psi_2(x)\}, \tag{4.2}$$

where $\psi(x)$ is a real two-component Fermi field.
Our notation for the lattice derivatives is

$$\Delta^+\varphi(x) = \varphi(x+1) - \varphi(x); \quad \Delta^-\varphi(x) = \varphi(x) - \varphi(x-1).$$

Now,

$$H = Q^2 = \sum_x \{\tfrac{1}{2}\pi^2(x) + \tfrac{1}{2}(\Delta^+\varphi)^2 + \tfrac{1}{2}V^2(\varphi)$$
$$-i\psi_1(x)\Delta^-\psi_2(x) + iV'(\varphi(x))\psi_1(x)\psi_2(x) + V(\varphi)\Delta^+\varphi(x)\}. \tag{4.3}$$

From eq. (4.3) we see that there is no bosonic doubling, and then—from Witten index arguments—we have no fermionic doubling either. Indeed, the dispersion relation is $E = \sin(K/2)$ $(-\pi \le K \le \pi)$ This means that there exists a Wilson term. In fact, we can rewrite the fermionic kinetic term as[2)]

$$\begin{aligned}-i\sum_x \psi_1 \Delta^- \psi_2 &= i\sum_x (\Delta^+\psi_1)\psi_2 = -i\sum_x \psi_2 \Delta^+ \psi_1 = \\ &= -\frac{i}{2}\sum_x [\psi_1 \Delta^a \psi_2 + \psi_2 \Delta^a \psi_1 - (\psi_1 \Delta^S \psi_2 - \psi_2 \Delta^S \psi_1)] = \\ &= -\frac{1}{2}[i\,\psi^T \alpha^1 \Delta^a \psi - \psi^T \gamma^0 \Delta^S \psi],\end{aligned} \tag{4.4}$$

where $(\Delta^a)\Delta^s$ is the (anti-)symmetric part of the lattice derivative[2)]

$$\begin{aligned}(\Delta^a \psi) &= \frac{1}{2}(\Delta^+ + \Delta^-)\psi(x) = \frac{1}{2}[\psi(x+1) - \psi(x-1)] \\ (\Delta^S \psi) &= \frac{1}{2}(\Delta^+ - \Delta^-)\psi(x) = \frac{1}{2}\Delta^+\Delta^-\psi(x).\end{aligned} \tag{4.5}$$

The term $i\psi^T\alpha^1\Delta^a\psi$ in eq. (4.4) is linear in p as $p \to 0$, where it gives the right continuum behavior and dominates the second term, which at $p \simeq \pi$ gives the dominant contribution (i.e. a mass term for the mirror fermions). I think that this automatic Wilson term is very interesting.

At this point, all the terms in H seem to have the correct continuum limit, because the last one tends to $V(\phi)\partial\phi/\partial x$ —a total derivative— in the naive continuum limit. In ref. 2 there are some remarks on the problem whether renormalization group effects can spoil the Lorentz covariance of the quantum continuum limit, due to this term. We recall here the argument because it will be useful for our considerations. For polynomial V the theory is super-renormalizable with dimension one coupling; the only terms induced by renormalization have dimension zero or one. Such terms are harmless being total derivatives.

Note that this argument relies on the fact that we have just one field; otherwise, the induced term would have the form

$$\sum_i F_i(\varphi)\frac{\partial\varphi}{\partial x^i}$$

which is not guaranteed to be a total derivative. We need the integrability conditions to be fulfilled by $F_i(\phi)$. In the next section we shall see how, for models having a stochastic structure (with two real fields) the integrability conditions will be automatically satisfied due to the peculiar structure of the associated LANGEVIN equation.

5. N=2 IN TWO DIMENSIONS WITH WILSON FERMIONS.

In order to obtain an N=2 lattice model with stochastic structure, the only thing we have to do is to discretize the effective 1D superpotential W of eq. (3.10). There are different, inequivalent, discretizations for this W which differ, for instance, for the fermionic kinetic term. In particular, we have discretizations with doubling and discretizations without doubling. In this section we discuss the model with Wilson fermions, in the next one the Susskind version.

In order to break chirality -as required to prevent doubling- we look for a discretization which explicitly breaks the complex structure of W, which corresponds to chirality. Writing

$$\varphi_x = 2^{-\frac{1}{2}}(A_x + iB_x)$$

we make the following Ansatz for W[13]

$$W[A_x, B_x] = \sum_x \{A_x[B_{x+1} - B_x] + 2\,\mathrm{Re}\,V[A_x + iB_x]\}\,. \tag{5.1}$$

Then, the bosonic Lagrangian becomes (see eq. (3.5))

$$\mathcal{L}_{BOS} = \frac{1}{2}\sum_x \left\{ \dot{A}_x + \dot{B}_x + (B_{x+1} - B_x)^2 + (A_{x+1} - A_x)^2 + 2\left(\frac{\partial \mathrm{Re} V}{\partial A_x}\right)^2 + 2\left(\frac{\partial \mathrm{Re} V}{\partial B_x}\right)^2 \right\}$$
$$+ \sum_x \left\{ (B_{x+1} - B_x)\frac{\partial \mathrm{Re} V}{\partial A_x} + (A_{x-1} - A_x)\frac{\partial \mathrm{Re} V}{\partial B_x} \right\}. \tag{5.2}$$

The Cauchy-Riemann equations allow us to rewrite the last term in eq. (5.2), the "spurious" term, as

$$(B_{x+1} - B_x)\frac{\partial \mathrm{Re} V}{\partial A_x} + (A_{x-1} - A_x)\frac{\partial \mathrm{Re} V}{\partial B_x} =$$
$$= (B_{x+1} - B_x)\frac{\partial \mathrm{Im} V}{\partial B_x} + (A_x - A_{x-1})\frac{\partial \mathrm{Im} V}{\partial A_x} \longrightarrow \partial_1 \mathrm{Im}\, V \tag{5.3}$$

in the naive continuum limit, and, indeed, if we work directly in the continuum, it is just this identity that makes the mapping to exist, in agreement with the algebraic argument of section 3.

The fermionic piece of the Lagrangian can be read in eq. (3.5)

$$\mathcal{L}_F = \bar{\xi}^A_x \dot{\xi}^A_x + \bar{\xi}^B_x \dot{\xi}^B_x + \bar{\xi}^A_x \Delta^+ \xi^B_x + \bar{\xi}^B_x \Delta^- \xi^A_x$$
$$+ 2\frac{\partial^2 \mathrm{Re} V}{\partial A_x^2}\left(\bar{\xi}^A_x \xi^A_x - \bar{\xi}^B_x \xi^B_x\right) + 2\frac{\partial^2 \mathrm{Re} V}{\partial A_x \partial B_x}\left(\bar{\xi}^A_x \xi^B_x + \bar{\xi}^B_x \xi^A_x\right). \tag{5.4}$$

The fermionic Hamiltonian is

$$H_F = \chi^+_\alpha(x)\left\{ i\sigma^2_{\alpha\beta}\Delta^a - \frac{1}{2}\sigma^1_{\alpha\beta}\Delta^+\Delta^- + 2\frac{\partial^2 \mathrm{Re} V}{\partial A_x^2}\sigma^1_{\alpha\beta}\right.$$
$$\left. + 2\frac{\partial^2 \mathrm{Re} V}{\partial A_x \partial B_x}\sigma^3_{\alpha\beta}\right\}\chi_\beta(x) \tag{5.5}$$

Just as in the previous case, the Hamiltonian (5.5) contains a Wilson term $-\frac{1}{2}\sigma^1_{\alpha\beta}\Delta^+\Delta^-$ to prevent the doubling and to make the fermionic dispersion relation to coincide with the bosonic one, as required by supersymmetry. (Let us note that this is not an ERS model with two ERS Majoranas.)

The niave continuum limit is OK.

At first sight, the spurious term may cause some problems in the continuum limit, but not very serious ones. The argument of ref. 2 works also in this case (let us remark, by the way, that our case is better because it is UV finite, not just super-renormalizable) except that we have to discuss the integrability conditions of the end of section 4.

The Langevin equation, eq. (3.6), corresponds to a Fokker-Planck equation(*)

$$-\partial_t \Psi = L^*_W \Psi \tag{5.6}$$

where

$$L^*_W = -\frac{1}{2}\sum_i \frac{\partial}{\partial x_i}\left[\frac{\partial}{\partial x_i} + 2\frac{\partial W}{\partial x_i}\right]. \tag{5.7}$$

From eq. (5.7) we see that the condition for having

$$L^*_W = L_{-W} \tag{5.8}$$

is

$$\frac{\partial}{\partial x_i}\frac{\partial}{\partial x_i} W = 0. \tag{5.9}$$

"Harmonicity" condition. $L^*_W = L_{-W}$ means that the backward Fokker-Planck equation (see ref. 17) is nothing else than the forward Fokker-Planck equation for the "backward" process of the end of section 3, which corresponds to the solution -W. The condition (5.8) corresponds to a peculiar symmetry for time-inversion. By studying how this symmetry is realized on the fermions, we see that—for the relevant models—it is a remainder of the original 4D PCT symmetry, and it is

(*) For a proof, see the lecture by B. Sakita in this volume. Difference of factors 2 is due to different normalization of the h-field.

related to the Hole Theory

$$H(\chi_\alpha \leftrightarrow \chi_\alpha^+, W \leftrightarrow -W) = H + \partial_i \partial_i W \tag{5.10}$$

It is obvious that eq. (5.9) is satisfied in our case, eq. (5.1) because of the analiticity of V. Using this peculiar symmetry, given that the derivative term of W -or, if you prefer of Q_A- is not renormalized, otherwise we would have dimension two induced operators, we can prove the integrability conditions. Indeed, the induced terms in W are ultralocal, and so can be written as a "superpotential" contribution to W, say F(A, B). The "harmonicity" condition becomes

$$\frac{\partial^2 F}{\partial A^2} + \frac{\partial^2 F}{\partial B^2} = 0 \tag{5.11}$$

i.e. F(A, B) = 2 ReV'(A+iB). This proves the existence of the necessary Cauchy-Riemann identities needed to cancel the spurious term in the continuum limit. For more than one N=2 supermultiplet, this proof of Lorentz covariance does not work; it is interesting, however, to notice that, in the general case, we have from eq. (5.9)

$$\sum_i \left\{ \frac{\partial^2 F}{\partial A_i^2} + \frac{\partial^2 F}{\partial B_i^2} \right\} = 0 \tag{5.12}$$

which is the finiteness condition[15].

Eq. (5.11) is -of course- just a consequence of the fact that we can do the same construction directly in the continuum case, with the same mapping; this suggests that we can work out the real space renormalization group directly on the mapping -or, if you prefer, on the stochastic process-. In ref. (10) we suggest that this fact is connected with the UV finiteness.

6. N=2 LATTICE SUSY: SUSSKIND FERMIONS.

These are essentially the Rittenberg-Yankielowicz models[18]. In a sense, they correspond to the most naive way to integrate -on a lattice- eq. (2.8), distinguishing even and odd sites with a "site grading" operator $(-)^x$. In this way we have

$$W(\varphi_x) = \sum_x \frac{(-1)^x}{2d}\varphi_x\varphi_{x+1} + \sum_x (-1)^x V(\varphi_x) \tag{6.1}$$

with φ_x a real field.

From the general algebraic argument, we expect N=2 SUSY —if Lorentz covariance has to be recovered—; in this case we need fermionic and bosonic doubling in order to reconstruct the whole N=2 supermultiplet. Indeed,

$$\begin{aligned} L = {} & \frac{1}{2}\sum_x \left\{ \dot\varphi_x^2 + \frac{1}{4d^2}(\varphi_{x+1}-\varphi_{x-1})^2 + \left(\frac{\partial V}{\partial \varphi_x}\right)^2 \right\} + \sum_x \frac{\partial V}{\partial \varphi_x}\frac{(\varphi_{x+1}-\varphi_{x-1})}{2d} + \sum_x \bar\xi_x \dot\xi_x \\ & + \sum_x \frac{(-1)^x}{2d}(\bar\xi_{x+1}-\bar\xi_{x-1})\xi_x + \sum_x (-1)^x \frac{\partial^2 V}{\partial \varphi_x^2}\bar\xi_x \xi_x \end{aligned} \tag{6.2}$$

with the redefinition of the fermionic fields

$$\phi_x = (i)^x (-1)^{[x/2]} \xi_x \, , \quad \phi_x^+ = (-i)^x (-)^{[x/2]} \xi_x \tag{6.3}$$

we can rewrite the fermionic piece of eq. (6.2) in the standard Susskind form[19]

$$\begin{aligned} L_F = {} & \sum_x \phi_x^+ \dot\phi_x + \frac{i}{2d}\sum_x (\phi_{x+1}^+\phi_x - \phi_x^+\phi_{x+1}) \\ & + \sum_x (-)^x \frac{\partial^2 V}{\partial \varphi_x^2}\phi_x^+\phi_x \, . \end{aligned} \tag{6.4}$$

The model is fine, except for the presence of the spurious term $\Sigma_x (\partial V/\partial \varphi_x)(\varphi_{x+1} - \varphi_{x-1})/2d$. Because the odd and even bosonic fields are different fields in the continuum limit, the spurious term does not have the structure of a total derivative.

This could be expected on general groupds. Indeed, from the continuous case and the Wilson example, we have learned that an expression of the form $(\partial V/\partial A)\partial B - (\partial V/\partial B)\partial A$ becomes a total derivative because we have a complex structure and hence Cauchy-Riemann equations. Then the relevant question is if and how the model can recover a complex structure in the quantum continuum limit. For fermions, the relevant complex structure is the same as the chiral structure. On a Susskind lattice, we have only discrete γ_5 symmetries -which corresponds to odd lattice translations- but it is generally believed that continuous chiral transformations could be recovered in the (quantum) continuum limit. In this case the complex structure would be also recovered. If the relevant Cauchy-Riemann identities would also be produced is an open question.

Then, the whole story boils down to understanding how continuous chirality is recovered in these Susskind models.

7. CONCLUSIONS AND DISCUSSION.

In this talk we have shown that one can construct N=2 two dimensional models which are equivalent to classical stochastic processes. For the models in eq. (5.2) and (5.5) the Langevin equations are

$$
\begin{aligned}
h_x^{(1)} &= A_x + B_{x+1} - B_x + 2\frac{\partial \operatorname{Re} V}{\partial A_x} \, , \\
h_x^{(2)} &= \dot{B}_x + A_{x-1} - A_x + 2\frac{\partial \operatorname{Re} V}{\partial B_x} \, .
\end{aligned}
\tag{7.1}
$$

The situation in four dimensions is rather tantalizing. We know that, if a N=2 Wess-Zumino model would exist it would be a stochastic model. But, apart from the free massless model, there seems to be none. In ref. 13 it was shown that some "pseudo-models" (a minimized version of the four dimensional models of ref. (18)) have a mapping. However, since they are Susskind models, they are expected not to have a covariant continuum limit because of the problem mentioned at the end of section 6. This corresponds to the fact that a Wilson model can be constructed (by a discretization of the continuum W) only for the free massless theory.

* * *

After the completion of this work, I received a paper by S. Elitzur and A. Schwimmer (ref. 20), where the same models are studied.

REFERENCES

(1) R. Haag, J.T. Lopuszanski and M. Sohnius, Nucl. Phys. B88 (1975) 257.

(2) S. Elitzur, E. Rabinovici and A. Schwimmer, Phys. Letters 119B (1982) 165.

(3) E. Witten, Nucl. Phys. B188 (1981) 513.

(4) E. Witten, Nucl. Phys. B202 (1982) 253.
S. Cecotti and L. Girardello Phys. Letters 110B (1982) 39.

(5) H.B. Nielsen and M. Nihomiya, Nucl. Phys. B185 (1981) 20; Nucl. Phys. B193 (1981) 173.

(6) K. Wilson, Erice Lectures (1975).

(7) H. Nicolai, Phys. Letters 89B (1980) 341. Nucl. Phys. B193 (1981) 173. But see also: M.L.L. Golterman "Comment on the proof of existence of the Nicolai mappings for scalar supersymmetric choices", preprint (1982).

(8) S. Cecotti and L. Girardello, Phys. Letters 110B (1982) 39.

(9) G. Parisi and N. Sourlas, Nucl Phys. B206 (1982) 321.

(10) S. Cecotti and L. Girardello, Annals of Physics 145 (1983) 81.

(11) F. Guerra, in "New Stochastic Methods in Physics", Physics Reports C77 (1981) No. 3.

(12) S. Cecotti and L. Girardello, Trieste ITCP preprint IC/82/105 (1982).

(13) S. Cecotti and L. Girardello, preprint CERN TH.3491 (1982).

(14) M. de Crombrugghe and V. Rittenberg, "Supersymmetric Quantum Mechanics", preprint BONN-HE-82-35.

(15) P.C. Davies, P. Salomonson and J.W. van Holten, Nucl. Phys. B208 (1982) 484.

(16) C. Itzykson and J.B. Zuber "Quantum Field Theory" McGraw-Hill (1980).

(17) C. De Witt-Morette and K.D. Elworthy in "New Stochastic Methods in Physics". Physics Report C77 (1981) No. 3.

(18) V. Rittenberg and S. Yankielowicz, CERN preprint TH.3263 (1982).

(19) L. Susskind, Phys. Rev. D16 (1977) 3031.

(20) S. Elitzur and A. Schwimmer "N=2 two dimensional Wess-Zumino Model on the lattice" preprint (1983).

NONLINEAR REALIZATION OF SUPERSYMMETRY

Julius Wess

Institut für Theoretische Physik
Universität Karlsruhe
D-75oo Karlsruhe 1, Germany

From chiral dynamics we know that low energy effects of a spontaneously broken symmetry can be well described by means of nonlinear realizations [1]. The Lagrangian, derived that way, should be considered as an effective Lagrangian at low energy. It exhibits the low-energy theorems in a model independent way. The lack of supersymmetry multiplets suggests that a similar approach should be tried for supersymmetric theories.

In this talk I will develop the formalism which Stuart Samuel will use in the next lecture to build realistic supersymmetric models.

The basic nonlinear transformation law was found by Akulov and Volkov [2] many years ago. Let $\tilde{\lambda}_\alpha(x)$ be a Weyl spinor [3] and ξ_α the parameter of a supersymmetry transformation, then it is easy to verify that the transformation law:

$$\delta_\xi \tilde{\lambda}_\alpha = \xi_\alpha - i(\tilde{\lambda}\sigma^m\bar{\xi} - \xi\sigma^m\bar{\tilde{\lambda}})\partial_m\tilde{\lambda}_\alpha \quad (1)$$

realizes a supersymmetry transformation, i.e.:

$$(\delta_\eta\delta_\xi - \delta_\xi\delta_\eta)\tilde{\lambda}_\alpha = -2i(\eta\sigma^m\bar{\xi} - \xi\sigma^m\bar{\eta})\partial_m\tilde{\lambda}_\alpha . \quad (2)$$

With the help of this intrinsically nonlinearly transforming field it is possible to realize supersymmetry transformation on any Lorentz covariant field:

$$\delta_\xi \tilde{C}_{Indesc} = -i(\tilde{\lambda}\sigma^m\bar{\xi} - \xi\sigma^m\bar{\tilde{\lambda}})\partial_m \tilde{C}_{Indesc} \quad (3)$$

and, again, it closes into a translation:

$$\left(\delta_\eta \delta_\xi - \delta_\xi \delta_\eta\right) \tilde{C}_{\text{Index}} = -2i(\eta\sigma^m\bar{\xi} - \xi\sigma^m\bar{\eta})\partial_m \tilde{C}_{\text{Index}} . \tag{4}$$

Index stands for any set of Lorentz or other indices. We shall call a field that transforms that way a "standard matter" field (SM field).

The nonlinear transformation law (1) of the A-V field can be derived as follows: Supersymmetry induces a motion in superspace [3]:

$$\begin{aligned} x'^m &= x^m + i(\theta\sigma^m\bar{\xi} - \xi\sigma^m\bar{\theta}) , \\ \theta'^\mu &= \theta^\mu + \xi^\mu , \qquad \bar{\theta}'_{\dot\mu} = \bar{\theta}_{\dot\mu} + \bar{\xi}_{\dot\mu} . \end{aligned} \tag{5}$$

If we consider a surface in superspace such that θ is a function of x, then (1) expresses the transformation of the surface unter the change of coordinates (5). (θ has been replaced by $\tilde{\lambda}$.) However, we know that supersymmetry can also be realized by the following motion [3]:

$$\begin{aligned} x'^m &= x^m + 2i\,\theta\sigma^m\bar{\xi} , \\ \theta'^\mu &= \theta^\mu + \xi^\mu , \qquad \bar{\theta}'_{\dot\mu} = \bar{\theta}_{\dot\mu} + \bar{\xi}_{\dot\mu} . \end{aligned} \tag{6}$$

The corresponding nonlinear transformation law is [4]:

$$\delta\lambda_\alpha = \xi_\alpha - 2i\lambda\sigma^m\bar{\xi}\partial_m\lambda_\alpha . \tag{7}$$

This is not an independent realization of supersymmetry, λ is a function of $\tilde{\lambda}$:

$$\lambda(x) = \tilde{\lambda}(y) \;, \quad y^m = x^m - i\,\tilde{\lambda}(y)\,\sigma^m\,\bar{\tilde{\lambda}}(y) \tag{8}$$

or, more explicit:

$$\lambda = \tilde{\lambda} - i\Big\{ \tilde{v}^m - i\,\tilde{v}^n \partial_n \tilde{v}^m - \tilde{v}^n \partial_n \tilde{v}^l \partial_l \tilde{v}^m - \frac{1}{2}\tilde{v}^n \tilde{v}^l \partial_n \partial_l \tilde{v}^m \Big\} \partial_m \tilde{\lambda} - \frac{1}{2}\tilde{v}^m \tilde{v}^n \partial_m \partial_n \tilde{\lambda} \tag{9}$$

with $\tilde{v}^m = \tilde{\lambda}\sigma^m \bar{\tilde{\lambda}}$.

Equation (9) has an inverse:

$$\tilde{\lambda} = \lambda + i\Big\{ v^m + i\,v^a(\partial_a \lambda \sigma^m \bar{\lambda} - \lambda \sigma^m \partial_a \bar{\lambda}) + v^a \partial_a v^b \partial_b v^m + \frac{1}{2} v^a v^b \partial_a \partial_b v^m \Big\} \partial_m \lambda - \frac{1}{2} v^a v^b \partial_a \partial_b \lambda \tag{10}$$

with $v^m = \lambda \sigma^m \bar{\lambda}$.

The transformation law (7) for λ is simpler because λ transforms into λ alone (and not into $\bar{\lambda}$ as well). The corresponding transformation law for a SM-field is:

$$\delta_\xi C_{Index} = 2i\lambda\sigma^m \bar{\xi} \partial_m C_{Index} . \qquad (11)$$

With any supersymmetry transformation we can associate a superfield. If the transformation law is such that

$$\delta_\xi a \equiv (\xi \delta + \bar{\xi}\bar{\delta}) \times a \qquad (12)$$

leads to

$$(\delta_\eta \delta_\xi - \delta_\xi \delta_\eta) a = -2i(\eta\sigma^m\bar{\xi} - \xi\sigma^m\bar{\eta})\partial_m a \qquad (13)$$

then it is possible to show [2] that

$$A(x,\theta,\bar{\theta}) = e^{\{\theta\delta + \bar{\theta}\bar{\delta}\}} \times a \qquad (14)$$

is a superfield.
The superfields corresponding to $\tilde{\lambda}$, λ and C_{Index} will be denoted by $\tilde{\Lambda}$, Λ and C_{Index} respectively. These superfields can also be defined through constraints. To derive these constraints, we make use of the identity:

$$D_\alpha e^{\{\theta\delta + \bar{\theta}\bar{\delta}\}} = e^{\{\theta\delta + \bar{\theta}\bar{\delta}\}} \delta_\alpha \qquad (15)$$

$$\bar{D}_{\dot{\alpha}} e^{\{\theta\delta + \bar{\theta}\bar{\delta}\}} = e^{\{\theta\delta + \bar{\theta}\bar{\delta}\}} \bar{\delta}_{\dot{\alpha}} .$$

This leads to the following constraint equations:

$$D_\beta \tilde{\Lambda}_\alpha = \varepsilon_{\alpha\beta} + i\,\sigma^m_{\beta\dot\rho}\,\bar{\tilde{\Lambda}}^{\dot\rho}\,\partial_m \tilde{\Lambda}_\alpha$$

$$\bar{D}_{\dot\beta} \tilde{\Lambda}_\alpha = -\,i\,\tilde{\Lambda}^\rho\,\sigma^m_{\rho\dot\beta}\,\partial_m \tilde{\Lambda}_\alpha$$

$$D_\beta \bar{\tilde{\Lambda}}_{\dot\alpha} = i\,\sigma^m_{\beta\dot\beta}\,\bar{\tilde{\Lambda}}^{\dot\beta}\,\partial_m \bar{\tilde{\Lambda}}_{\dot\alpha} \tag{16}$$

$$\bar{D}_{\dot\beta} \bar{\tilde{\Lambda}}_{\dot\alpha} = -\,\varepsilon_{\dot\alpha\dot\beta} - i\,\tilde{\Lambda}^\beta\,\sigma^m_{\beta\dot\beta}\,\partial_m \bar{\tilde{\Lambda}}_{\dot\alpha}$$

$$D_\beta\, \tilde{C}_{Ind} = i\,\sigma^m_{\beta\dot\rho}\,\bar{\tilde{\Lambda}}^{\dot\rho}\,\partial_m \tilde{C}_{Ind}$$

$$\bar{D}_{\dot\beta}\, \tilde{C}_{Ind} = -\,i\,\tilde{\Lambda}^\rho\,\sigma^m_{\rho\dot\beta}\,\partial_m \tilde{C}_{Ind}\,.$$

The constraints for the new superfields Λ, $\bar{\Lambda}$ and C_{Index} look simpler:

$$D_\beta \Lambda_\alpha = \varepsilon_{\alpha\beta} \tag{17}$$

$$\bar{D}_{\dot\beta} \Lambda_\alpha = -2i\,\Lambda^\rho\,\sigma^m_{\rho\dot\beta}\,\partial_m \Lambda_\alpha$$

$$D_\beta \bar{\Lambda}_{\dot\alpha} = 2i\,\sigma^m_{\beta\dot\beta}\,\bar{\Lambda}^{\dot\beta}\,\partial_m \bar{\Lambda}_{\dot\alpha}$$

$$\bar{D}_{\dot\beta} \bar{\Lambda}_{\dot\alpha} = -\,\varepsilon_{\dot\alpha\dot\beta}$$

$$\tag{17}$$

$$D_\beta\, C_{Ind} = 0$$

$$\bar{D}_{\dot\beta}\, C_{Ind} = -2i\,\Lambda^\gamma\,\sigma^m_{\gamma\dot\beta}\,\partial_m C_{Ind} \;.$$

These constraint equations define a superfield. If we solve them we would find that all the higher components of these superfields are functions of the lowest component and its derivatives. It is exactly the same superfield as it was defined through equ. (14).

It is possible to decompose a superfield into SM-fields and, inversely, to build a superfield from SM-fields. This construction was given by Ivanov and Kapustnikov [5] . Here I would like to describe a slightly different construction which is easier to be generalized to curved space: Let me demonstrate the method on a chiral field ϕ : $\bar{D}_{\dot\alpha}\phi = 0$. There is a function of the chiral field ϕ and the A-V superfield Λ which satisfies the constraints of a SM superfield:

$$C = \phi - \Lambda^\alpha D_\alpha \phi - \frac{1}{4}\Lambda^2 D^2 \phi \tag{18}$$

this can be verified directly, using $\bar{D}_{\dot\alpha}\phi = 0$ and equs. (17). The lowest component of A is a function of the components of ϕ and of λ :

$$\phi = a + \sqrt{2}\,\theta\psi + \theta\theta f \tag{19}$$

$$C\big|_{\theta=\bar\theta=0} \equiv c = a - \sqrt{2}\,\psi\lambda + f\lambda^2 .$$

The component field c transforms in agreement with (11) and all the higher components of C are functions of λ and c and their derivatives.

The construction (18) works for a chiral superfield. To find the SM field that corresponds to the spinor component of ϕ we have to start from a chiral superfield which has ψ as a lowest component. Such a superfield is

$$\Psi^{(0)}_\alpha = -\frac{1}{4}\bar D^2\left(\bar\Lambda^2 D_\alpha \phi\right) . \tag{20}$$

It is chiral and because

$$-\frac{1}{4}\bar D^2\bar\Lambda^2 = 1 + \cdots \tag{21}$$

$\Psi^{(0)}_\alpha$ starts with $D_\alpha\phi$.

The SM superfield corresponding to Ψ^0_α is:

$$\Gamma_\alpha = \Psi^{(0)}_\alpha - \Lambda^\beta D_\beta \Psi^{(0)}_\alpha - \frac{1}{4}\Lambda^2 D^2 \Psi^{(0)}_\alpha \tag{22}$$

its lowest component γ_α is a SM field which corresponds to ψ_α.

Similarly:

$$\mathcal{F}^{(0)} = \frac{1}{16} \bar{D}^2 (\bar{\Lambda}^2 D^2 \phi) \tag{23}$$

is a chiral superfield which has f in its lowest component.

$$F = \mathcal{F}^{(0)} - \Lambda^\alpha D_\alpha \mathcal{F}^{(0)} - \frac{1}{4} \Lambda^2 D^2 \mathcal{F}^{(0)} \tag{24}$$

is the SM superfield corresponding to f. The corresponding SM field (the lowest component of F) we **denote** by φ.

We have demonstrated that it is possible to change the field variables a, ψ and f to new variables c, γ_α and φ which transform like SM fields. This is an allowd change of interpolating fields such that the physics should remain unchanged.

The decomposition of a chiral superfield into SM fields has an inverse. Assume that there is a set of SM fields c, γ_α and φ. Their supersymmetry transformation law is given through (11).

The corresponding SM superfield can be constructed according to (14).

$$\begin{aligned} C &= e^{\{\theta \mathcal{J} + \bar{\theta} \bar{\mathcal{J}}\}}{}_\times c \\ \Gamma_\alpha &= e^{\{\theta \mathcal{J} + \bar{\theta} \bar{\mathcal{J}}\}}{}_\times \sqrt{2}\, \gamma_\alpha \\ F &= e^{\{\theta \mathcal{J} + \bar{\theta} \bar{\mathcal{J}}\}}{}_\times \varphi \quad . \end{aligned} \tag{25}$$

A chiral projection can be applied to these superfields as well as to :

$$\phi^{(0)} = -\frac{1}{4}\bar{D}^2(\bar{\Lambda}^2 C)$$

$$\Gamma^{(0)}_\alpha = -\frac{1}{4}\bar{D}^2(\bar{\Lambda}^2 \Gamma_\alpha) \tag{26}$$

$$F^{(0)} = -\frac{1}{4}\bar{D}^2(\bar{\Lambda}^2 F)$$

$$\Lambda^{(0)}_\alpha = -\frac{1}{4}\bar{D}^2(\bar{\Lambda}^2 \Lambda_\alpha) .$$

These chiral superfields can be combined to a chiral superfield with the usual number of independent component fields:

$$\phi = \phi^{(0)} + \Lambda^{(0)\alpha}\Gamma^{(0)}_\alpha + \Lambda^{(0)2} F^{(0)} . \tag{27}$$

This is the inverse of the decomposition of a chiral superfield into SM fields. If we would have decomposed ϕ into the SM fields C, γ_α and f and then gone through the constructions outlined above, then (27) would be an identity.

Note that ϕ of equ. (27) remains a chiral superfield even if we would have put $\phi^{(0)}$, or $\Gamma^{(0)}$, or $F^{(0)}$ equal to zero. This suggests how certain components of a superfield can be eliminated in a "supersymmetric way". In the next lecture, S. Samuel will make use of this possibility with the purpose to construct supersymmetric Lagrangians without partner fields. A general prescription how to build supersymmetric Lagrangians from SM fields has been given in[6].

REFERENCES

1) S. Coleman, J. Wess and B. Zumino, Phys. Rev. 177 (1969) 2239;
C. Callan, S. Coleman, J. Wess and B. Zumino, Phys. Rev. 177 (1969) 2247;
J. Wess, Springer Tracts in Modern Physics, Vol. 50 (1969) 132.

2) V. Akulov and D. Volkov, Phys. Lett. 46B (1973) 109;
D. Volkov and V. Soroka, JETP Lett. 18 (1973) 312.

3) J. Bagger and J. Wess, Supersymmetry and Supergravity, Princeton Series in Physics, Princeton University Press (1983).

4) S. Samuel and J. Wess, Columbia University preprints (CU-TP-260, CU-TP-258), to be published in Nuclear Physics.

5) E. Ivanov and A. Kapustnikov, J. Phys. A11 (1978) 2375; J. Phys. 68 (1982) 167.

6) J. Wess, Nonlinear Realization of the N=1 Supersymmetry, University of Karlsruhe preprint (1982).

SECRET SUPERSYMMETRY MODEL BUILDING

Stuart Samuel

Department of Physics
Columbia University
New York, New York 10027

ABSTRACT

A new model building technique is introduced in the context of non-linear supergravity.

1. Introduction

I will talk about work done with Julius Wess under the title of Secret Supersymmetry which is now a Columbia University preprint.[1] Our work is based on the papers of Akulov and Volkov[2] and Ivanov and Kapustnikov.[3] A complete set of references can be found in ref. 1.

We have constructed an interesting class of realisitic supersymmetric models using non-linear realizations of supersymmetry. The interesting feature of our models is that particles do not necessarily have supersymmetric partners. When one builds a model using linear representations of supersymmetry, particles appear in supersymmetric multiplets. The electron is always accompagnied by a scalar partner (the selectron), the photon has a fermionic brother (the photino), etc. If supersymmetry is spontaneously broken, broken multiplets appear, that is, there are mass splittings between members of a multiplet. One challenge in realistic model building is to make the right members heavy and experimentally unobservable. In our case, this is not a problem; there simply are no partners.

In this talk I will present a QED example. We will construct a supersymmetric model of QED in which the basic quanta are the photon, the electron, a heavy gravitino, and the graviton. The gravitino and graviton appear because supergravity is a necessary ingredient in our construction. What is important is that there is no scalar electron. When the effects of gravity are ignored, the model becomes the familiar one for QED (see Eq.(24) below).

We have also created a model which at low energies reduces to the standard Weinberg-Salam model (except for a slightly more complicated Higgs sector). It possesses no unobserved low mass particles. In fact, the only quanta, not yet experimentally observed, are the Higgs scalars, a massive gravitino, and the graviton. There are no phenomenological problems: The masses of all particles can be adjusted to their experimental values, the cosmological constant can be arranged to be zero, all weak interaction phenomenology is reproduced, including the absence of flavor changing neutral currents, etc. The fact that the original theory was supersymmetric is well hidden.

Given that no (broken) supersymmetric multiplets have yet been observed in nature, one might speculate that nature realizes supersymmetry in our non-linear mode. It may be that nature has an underlying symmetric structure which is difficult to detect. The same would be true in a linear model if all the partner particles could be made to be very heavy. Are there, then, any low energy signals? In our QED-like example there are none. In our Weinberg-Salam-like model, the situation is experimentally somewhat more promising. All realistic supersymmetric models require at least two sets of Higgs doublets, twice the number as in the standard Weinberg-Salam model. The reason for this is simple: The basic representations of supersymmetry involve fermions of definite chirality. The corresponding scalar partners inherit this chirality, so that in supersymmetry there is a notion of a "right-handed" (or "left-handed") scalar. This leads to restrictions in the form of the couplings. Recall that in the standard Weinberg-Salam model, the Higgs doublet couples to the left-handed quark doublet and the right-handed anti down-quarks. It also couples to the left-handed lepton doublet and the right-handed positively charged leptons. When the neutral component acquires a vacuum expectation value, the charged leptons and down quarks get masses. To generate up quark masses, an SU(2) conjugate Higgs field is constructed from the complex conjugate of the Higgs. It transforms as an SU(2) doublet but has opposite weak hypercharge. It couples to the left-handed quark doublet and the right-handed anti up quarks. Due to the restrictive nature of supersymmetry couplings, the complex conjugated Higgs field cannot couple to the right-handed anti up quark. Hence in supersymmetry, one must replace the SU(2) conjugated Higgs field by a second Higgs doublet. Although our model does not have fermionic partners for the Higgs fields, the Higgs fields retain their sense of chirality and hence two Higgs fields are required no matter what model building technique is used. Therefore, if nature realizes supersymmetry in our non-linear mode or if nature realizes supersymmetry in the standard (linear) mode but all supersymmetric partners get very heavy (?"super" heavy) then the only low energy experimental test of supersymmetry will be the number of Higgs particles and the way they couple to quark and lepton matter. One Higgs should couple only to the right-handed leptons and right-handed down quarks; the other Higgs scalar should only couple to the right-handed up quarks. From an experimental viewpoint, the Higgs sector might be the only way of sensing an underlying supersymmetric structure in nature.

Before discussing our model building technique, I would like to remind you of a few elementary facts about supersymmetry. Given any positive energy eigenstate of definite fermion number, a state of the same energy but opposite fermion character can be constructed by applying the supersymmetric charges, Q_α, to the original state. This follows because the Q_α's carry fermion number and commute with the Hamiltonian. There is always degeneracy among fermionic and bosonic energy eigenstates. When the vacuum is annihilated by the Q_α's, that is, when supersymmetry is not spontaneously broken, there must be mass degeneracy among bosons and fermions. Since no such degeneracy is observed in nature, supersymmetry must necessarily be spontaneously broken. The outstanding phenomenological problem is to find viable and realistic versions of spontaneously broken supersymmetry theories. When supersymmetry is spontaneously broken, a (massless) Goldstone fermion appears, the fermionic analog of the Goldstone boson. The Q_α's are then ill-defined operators when acting on the Hilbert space of physical states. In a non-rigorous sense, the degeneracy between boson and fermionic states is still maintained. To any given energy eigenstate of definite fermion number, one can add a massless goldstino in the limit in which its momentum goes to zero. Such a state will have the same energy and momentum as the original state but opposite fermion character. This is an intuitive way of understanding the spontaneously broken case. In our model building technique, we will be realizing supersymmetry in this manner.

There is another phenomenon that we will need, and it is the fermionic analog of the Higgs mechanism. When supersymmetry is made local, that is we go to supergravity, the Goldstone fermion associated with spontaneously broken supersymmetry is eaten by the gravitino (the spin 3/2 fermionic partner of the spin 2 graviton) and becomes massive. The +1/2 and -1/2 helicity states of the goldstine combine with the +3/2 and -3/2 helicity states of the gravitino to yields one massive spin 3/2 particle. This is the superhiggs mechanism in supergravity. This phenomenon will be an important ingrediant in our model building technique. It is, in general, quite useful since it allows one to trade two unobserved massless particles for one heavy (unobserved) one.

A word about notation. I will use the notation in ref. 4 which uses two-component spinors. When I write Q_α and $\bar{Q}_{\dot\alpha}$ the indices α and $\dot\alpha$ will only take on the values 1 and 2. If you are use to working with four component objects, you can arrange these into a four-component (Majorana) object via

$$\begin{pmatrix} Q_\alpha \\ Q^{\dot\alpha} \end{pmatrix} \quad . \tag{1}$$

I will make use of superfields which is a very useful device for grouping the representations of supersymmetry into a single object. One introduces two (constant) anticommuting parameters θ_α and $\bar\theta_{\dot\alpha}$. A general superfield is a polynomial expression in θ and $\bar\theta$ whose coefficients are the component fields:

$$\Phi(x,\theta,\bar\theta) = A(x) + \sqrt{2}\,\theta\psi(x) + \ldots, \tag{2}$$

Here, $A(x)$ is a boson field, $\psi(x)$ is a fermion field, etc. I will also frequently introduce addition anticommuting variables to make notation index free. Examples are

$$\begin{aligned} Q\,\xi &\equiv Q^\alpha \xi_\alpha \;, \\ \eta\,\lambda(x) &\equiv \eta^\alpha \lambda_\alpha(x), \end{aligned} \tag{3}$$

where ξ_α and η^α are constant anticommuting varibles.

A general superfield does not correspond to an irreducible representation of supersymmetry. One must impose constraints to limit the number of component fields. Examples are the chiral superfield:

$$\bar D_{\dot\alpha}\,\Phi = 0, \tag{4}$$

where

$$\begin{aligned} \bar D_{\dot\alpha} &= -\frac{\partial}{\partial\bar\theta^{\dot\alpha}} - i\theta^\alpha \sigma^m_{\alpha\dot\alpha}\,\partial_m \;, \\ D_\alpha &= \frac{\partial}{\partial\theta^\alpha} + i\sigma^m_{\alpha\dot\alpha}\,\bar\theta^{\dot\alpha}\,\partial_m \;, \end{aligned} \tag{5}$$

and the real or vector superfield:

$$V = V^\dagger \;. \tag{6}$$

The chiral superfield contains a Majorana fermion and a complex scalar. The vector superfield contains a spin one particle (a photon, for example) and a spin 1/2 Majoran fermion (the photino, for example):

$$V = \ldots - \theta\sigma^m\bar{\theta}v_m - i\bar{\theta}\bar{\theta}\theta\lambda + \ldots \quad (7)$$

2. The Model Building Technique

Non-linear representations of supersymmetry were presented in the talk by Julius Wess and can also be found in our paper.[1] A simple example of a non-linear representation of flat space supersymmetry is[2]

$$\delta\lambda_\alpha \equiv (\xi Q + \bar{\xi}\bar{Q}) \times \lambda_\alpha = \frac{1}{k}\xi_\alpha - 2ik(\lambda\sigma^m\bar{\xi})\,\partial_m\lambda_\alpha . \quad (8)$$

ζ and $\bar{\xi}$ are the anticommutting parameters associated with the supersymmetry transformation, k is an adjustable parameter of dimensions $(\text{mass})^{-2}$. $\lambda(x)$ is the non-linearly transforming field. What one must verify is that the anticommutator algebra of the Q's is satisfied:

$$[\xi'Q + \bar{\xi}'\bar{Q}, \xi Q + \bar{\xi}\bar{Q}] = -2i\,(\xi'\sigma^m\bar{\xi} - \xi\sigma^m\bar{\xi}')\,\partial_m . \quad (9)$$

The transformation law in Eq.(8) can be regarded as describing goldstino dynamics. The corresponding theory is the supersymmetric version of the π and σ model.

There is a superfield, which I will denote by Λ_α, corresponding to Eq.(8):

$$\begin{aligned}\Lambda\eta = \lambda\eta &+ \frac{\theta\eta}{k} - \underline{2ik\,\lambda\sigma^m\bar{\theta}\partial_m\lambda\eta} - i\theta\sigma^m\bar{\theta}\partial_m\lambda\eta \\ &- 4k^2\lambda\sigma^n\bar{\theta}\partial_n\lambda\sigma^m\bar{\theta}\partial_m\lambda\eta - 2k^2\lambda\sigma^n\bar{\theta}\lambda\sigma^m\bar{\theta}\partial_n\partial_m\lambda\eta \\ &\underline{- 2k\,\theta\sigma^m\bar{\theta}\partial_m(\lambda\sigma^n\bar{\theta}\partial_n\lambda\eta)} - \frac{1}{2}\theta\sigma^m\bar{\theta}\theta\sigma^n\bar{\theta}\partial_m\partial_n\lambda\eta\end{aligned} \quad (10)$$

The coefficients of the θ's are the field operators associated with this representation of supersymmetry. The lowest component of Λ_α is λ_α, the original non-linear (goldstino) field. All the other partner operators are composite in character. In Eq.(10), I have underlined several terms. These are bosonic operators constructed from an even number of λ's. Unlike (linear) representations of supersymmetry where the partner operators of fermions are new bosonic fields, the partner operators of λ are composite.

A supersymmetric Lagrangian for λ is[2]

$$\mathcal{L} = -\frac{k^2}{2}\int d^2\theta d^2\bar{\theta}(\Lambda\Lambda)(\bar{\Lambda}\bar{\Lambda})$$

$$= -\frac{1}{2k^2} + \frac{i}{2}(\partial_m\lambda\sigma^m\bar{\lambda} - \lambda\sigma^m\partial_m\bar{\lambda}) \qquad (11)$$

$$+ \text{ interaction terms.}$$

The first term in Eq.(11) is a constant background energy density. This signals that supersymmetry is spontaneously broken. Theories involving always realize supersymmetry in the spontaneously broken mode. The λ is massless and is the goldstino.

Here is the basic idea behind our model building technique. Suppose we wish to construct a theory containing certain matter fields. To make the theory supersymmetric, normally one would have to introduce all the supersymmetric partners. Instead, we will introduce only the λ field. The supersymmetric partner operators will be composite operators constructed out of the basic matter field and the goldstino. These operators will produce a basic matter field and a λ particle. The boson-fermion energy degeneracy will be realized as discussed above, that is, via the addition of a momentumless goldstino. Several theoretical challenges must be overcome to implement such a program. First, one must show that such representations exist. Are there superfields involving a basic matter field and composite partners? Secondly, one must confront the problem of renormalizability. Lagrangians involving non-linear superfields are always non-renormalizable. Typically many powers of λ enter in the interactions. The way to overcome this is to note that these non-renormalizable terms would be gauge degrees of freedom if the supersymmetry were made local. If the theory can be embbedded in a supergravity theory, then the non-renormalizable terms are simply gauge artifacts and can be gauged away. The third challenge is to implement this and find the corresponding non-linear representations in curved space. We have succeeded in doing this.

The most tedious aspect of our work was to find the non-linear representations of supersymmetry in the supergravity case. This involved an enormous number of hours of guesswork and calculation. Finally, we were able to find them. Julius presented these new representations of the supergravity algebra in his talk and they can be found in our paper.[1)]

It should be mentioned that our models are still not renormalizable. This is not because of the use of non-linear representations of supersymmetry but because we have had to introduce gravity. Presently, there are no known renormalizable theories of gravity and so our models are non-renormalizable at the Planck scale. When gravitational effects are turned off, only renormalizable interaction terms remain. Some day, someone may find a renormalizable supergravity theory and it will be interesting to see if we can adapt our methods to that theory.

Summarizing, to build a supersymmetric model without partner fields, you must make use of the non-linear λ field. You must intertwine it with the basic matter fields to form composite supersymmetric partner operators. Then you must make the model locally supersymmetric in order to obtain a low-energy renormalizable theory. The physics can be analyzed in a gauge where the gravitino manifestly eats the goldstino field. At the same time, it eats the non-renormalizable interactions and the composite operators.

The superfield in Eq.(10) can be defined via constraints:

$$\begin{aligned} \xi D \Lambda\eta &= \frac{1}{k}\,\xi\eta \ , \\ \bar{\xi}\bar{D}\Lambda\eta &= -\,2ik(\Lambda\sigma^m\bar{\xi})\,\partial_m\Lambda\eta, \\ \Lambda\eta| \equiv \Lambda\eta|_{\substack{\theta=0\\ \bar{\theta}=0}} &= \lambda\eta \ . \end{aligned} \tag{12}$$

Eqs.(12) are completely equivalent to the expression in Eq.(10). To go to the supergravity case, one needs to find the correct generalization in curved space. They turn out to be[5)]

$$\begin{aligned} \bar{\zeta}\bar{\mathcal{D}}\,\Lambda\eta &= -\,2ik\,\Lambda\,\sigma^a\bar{\zeta}\,\mathcal{D}_a\,\Lambda\eta + \frac{k}{2}\,\Lambda\Lambda\bar{\zeta}G\eta, \\ \zeta\mathcal{D}\,\Lambda\eta &= \zeta\eta\left(\frac{1}{k} - 2kR^*\Lambda\Lambda\right), \\ \Lambda\eta\,| &= \lambda\eta \ . \end{aligned} \tag{13}$$

$\mathcal{D}_\alpha$ and $\bar{\mathcal{D}}_{\dot{\alpha}}$ are the curved space generalizations of D_α and $\bar{D}_{\dot{\alpha}}$. Here R and G_a are supergravity superfields (see, for example, ref. 4). The reason for their appearance on the right-hand side of Eq.(13) is due to the fact that operator algebra of $\mathcal{D}_\alpha$, $\bar{\mathcal{D}}_{\dot{\alpha}}$, and $\mathcal{D}_a$ (the curved space analog of ∂_a) is more complicated in the curved space case.

Let me now present an example of a non-linear representation of supersymmetry involving an ordinary matter field and the non-linear field:[3)]

$$\delta a_{index} = 2ik(\xi\sigma^m\bar{\lambda})\partial_m a_{index} \quad . \tag{14}$$

Here index stands for any Lorentz index, a μ , a $\dot{\mu}$, an m, etc. Since does not appear in Eq.(14), the corresponding superfield, $A_{index}(x)$, is chiral:

$$A = a(y) + 2ik(\theta\sigma^m\bar{\lambda})\partial_m a(y)$$

$$+ (-2k^2(\bar{\lambda}\bar{\sigma}^m\sigma^n\partial_m\bar{\lambda})\partial_n a + k^2(\bar{\lambda}\bar{\lambda})\partial^n\partial_n a)\theta\theta, \tag{15}$$

$$y^m = x^m + i\,\theta\sigma^m\bar{\theta} \ .$$

Compare this to an ordinary scalar chiral superfield

$$\Phi = a(y) + \sqrt{2}\ \theta\psi(y) + \theta\theta\ f(y). \tag{16}$$

Notice that the fermionic partner operator is an operator bilinear in λ and a(x). Even the auxilary "f" field is composite. In short, there exist non-linear realizations of supersymmetry involing ordinary matter fields and the goldstino field. The corresponding constraints have been found in the supergravity case, and these can be found in Julius's talk or in our paper (Eqs.(2.14), (2.16), (2.18), (2.20), (2.22), and (2.23)).[1)] These constraints involve the gravitational superfields R^*, $G_{\alpha\dot{\alpha}}$, and $W_{\alpha\beta\gamma}$. Physically, this means that the composite operators will now involve not only the λ field but also gravitational fields, that is, the graviton and the gravitino.

Let me now illustrate our technique with a QED model. In our paper, we have presented a step by step procedure. First, one must begin with a flat space supersymmetric version of QED:

$$\mathcal{L}_{QED} = \int d^2\theta d^2\theta\{E_R^* \exp(2eV)\ E_R^{*\dagger} + E_L^{\dagger} \exp(-2eV)\ E_L\}$$

$$+ (\int d^2\theta\{\tfrac{1}{4}\ W^\alpha W_\alpha + m_e E_R^* E_L\} + h.c). \tag{17}$$

Here E_L and E_R^* are chiral superfields which contain the left-handed components of the electron fields and the right-handed components of the positron field. m_e and e are the mass and charge of the electron. V is a vector superfield which contains the photon and $W_\alpha = -\frac{1}{4}\bar{D}\bar{D}D_\alpha V$. If E_L, E_R^*, and V were ordinary (linear) superfields then the theory would also contain the supersymmetric partners of the electron (the selectron) and the photon (the photino). In curved space, the corresponding Lagrangian is

$$\mathcal{L}_{QED} = \int d^2\Theta 2\mathscr{E} \{ \frac{1}{4} WW + m_e E_R^* E_L$$

$$- \frac{1}{8}(\bar{\mathscr{D}}\bar{\mathscr{D}} - 8R) [E_R^* \exp(2eV) E_R^{*\dagger} \tag{18}$$

$$+ E_L^\dagger \exp(-2eV) E_L] \} + h.c.,$$

to which the Lagrangian of supergravity and the λ field must be added

$$\mathcal{L}_{gravity + non-linear} = \int d^2\Theta 2\mathscr{E} \; (- \frac{3R}{\kappa^2} + \Delta + \frac{3\phi'}{\sqrt{2}\,k}) \tag{19}$$

$$+ h.c.$$

In Eq.(19), $2\mathscr{E}$ is the supersymmetric version of the determinant of the metric, R, the same superfield mentioned above, is a supersymmetric version of the scalar curvature, κ is related to Newton's constant, and Δ is a supersymmetric version of the cosmological constant. It generates a mass for the gravitino as well as a negative constant background energy density. The last term in Eq.(19) is the goldstino action which produces the kinetic energy term of the goldstino, self-interaction terms, as well as a positive background energy density. The two energy densities can be adjusted by hand to cancel. When this is done, the cosmological constant is zero and one gets the standard relation among the gravitino mass, $m_{3/2}$, the Planck mass, m_P, and the parameter k:

$$m_{3/2}\, m_P = \frac{8\sqrt{2\pi}}{k} \quad . \tag{20}$$

$1/\sqrt{k}$ can be regarded as the scale of supersymmetry breaking so the Eq.(20) says that the supersymmetry breaking scale is the geometric average of the Planck mass and the gravitino mass.

Instead of using ordinary (linear) superfields, use non-linear superfields for E_L, E_R^*, and V in Eq.(18):

$$E_R^* = \sqrt{2}\, k \Lambda_o \Psi_R \;,$$

$$E_L = \sqrt{2}\, k \Lambda_o \Psi_L \;,$$

$$V = \ldots - k^2 \Lambda\sigma^a\bar{\Lambda} A_a + k^4(\Lambda\Lambda\bar{\Lambda}\bar{\Lambda})\, G^a A_a \qquad (21)$$

$$- k^6 (\Lambda\Lambda\bar{\Lambda}\bar{\Lambda})(\mathcal{D}_a \Lambda\sigma^b \mathcal{D}^a\bar{\Lambda})\, A_b ,$$

$$W_\alpha \equiv -\frac{1}{4}(\bar{\mathcal{D}}\bar{\mathcal{D}} - 8R)\,\mathcal{D}_\alpha V \;.$$

A_a is a real vector non-linear superfield (see Eq.(2.20) of our paper). We have only shown the important terms in V; there are other terms present which are U(1) gauge degrees of freedom (see Eq.(3.13) in our paper). Ψ_L and Ψ_R are chiral non-linear superfields involving the electron field (see Eq.(2.16) of our paper), Λ_o is a curved space chiral version of the Λ superfield:

$$\Lambda_o\, \eta \equiv -\frac{1}{4}(\bar{\mathcal{D}}\bar{\mathcal{D}} - 8R)(k^2\bar{\Lambda}\bar{\Lambda}\Lambda\eta). \qquad (22)$$

We have replaced V by a linear combination of superfields which is still a real vector superfield in the sense that it is equal to its complex conjugate. E_L and E_R^* are the product of two chiral superfields and hence chiral. Because chiral (non-linear) superfields are substituted for (linear) chiral superfields and real (non-linear) superfields are substituted for real (linear) superfields, the local supersymmetry of the model is maintained.

Our supersymmetric theory of QED is the sum of the two Lagrangians in Eqs.(18) and (19). In an arbitrary gauge it looks quite complicated. In our paper (Sec.IV), we have given a method for obtaining the component Lagrangian in an arbitrary gauge. There is one gauge, however, which is more physical than the others, in which the λ field is gauged away. In this gauge the gravitino has manifestly eaten the goldstino. Let

$$\psi_e = \begin{pmatrix} \psi_{L\alpha} \\ \\ \bar{\psi}_R^{\dot{\alpha}} \end{pmatrix} , \tag{23}$$

be the four-component Dirac electron. Then the Lagrangian in the unitary gauge is

$$\mathcal{L} \simeq -\frac{1}{4} F_{\mu\nu} F^{\mu\nu} + \bar{\psi}_e (i\gamma^\mu \partial_\mu - e\gamma^\mu A_\mu - m_e)\psi_e \tag{24}$$

$$+ \text{ gravitational interactions.}$$

The gravitational interactions can be found in Eq.(4.5) of our paper. The couplings of the electron to the gravitino are of the four-fermion type and proportional to Newton's constant.

Both the gravitino mass and the Planck mass are parameters under our control which we can make large. In this limit, Eq.(24) is the low energy Lagrangian. The underlying theory is supersymmetric but at low energies one would never know it. We have thus constructed a model of QED in which neither the electron nor the photon has a supersymmetric partner.

In this talk I have presented a new interesting model building technique. I have constructed a supersymmetric model of QED in which neither the electron nor the photon has a supersymmetric partner. The physical particle excitations were a massive electron, a massless photon, a heavy gravitino, and a massless graviton. Supersymmetry was realized in the spontaneously broken mode in which the goldstino was used to achieve degeneracy among fermionic and bosonic energy eigenstates. In general, we can build models in which for each fermion there need not be a boson (and vice versa) even though the underlying Lagrangian is supersymmetric. We accomplished this by using non-linear supergravity. The partner operators were composite operators constructed from a basic matter field, the goldstino fermion, and gravitational quanta. These partner operators did not produce a new particle state of opposite fermion character, rather they produced a multiparticle state consisting of a basic particle and gravitational excitations.

In our paper, we have also built an $SU(3)\times SU(2)\times U(1)$ model which incorporates all four forces of nature. In the low energy limit, the theory reduces to the standard Weinberg-Salam model except there is an additional Higgs doublet. The couplings of the Higgs fields to quark and lepton matter is restricted in a way characteristic of supersymmetric theories. This low energy Lagrangian is renormalizable so that it could, for example, be analyzed beyond the tree level. It reproduces all the current known low energy phenomenology. It is an example which demonstrates that supersymmetry cannot be ruled out on phenomenological grounds. Although the theory is supersymmetric, supersymmetry manifests itself in a very subtle manner. Without partner particle states, the likely experimental indications of supersymmetry will come only through the charcteristic couplings mentioned above. In short, that nature is supersymmetric made be well hidden.

REFERENCES

1. S. Samuel and J. Wess, Secret Supersymmetry, Columbia University preprint CU-TP-264 (June, 1983).

2. V. P. Akulov and D. V. Volkov, Phys. Lett. 46B (1973) 109.

3. E. Ivanov and A Kapustnikov, J. Phys. All (1978) 2375; J. Phys. G8 (1982) 167.

4. J. Wess and J. Bagger, Supersymmetry and Supergravity, (Princeton University Press, 1983).

5. S. Samuel and J. Wess, Nucl. Phys. B221 (1983) 153.

N=4 YANG-MILLS THEORY
A UV FINITE QUANTUM FIELD THEORY

by

Lars Brink
Institute of Theoretical Physics
S-412 96 Göteborg
Sweden

ABSTRACT

The N=4 supersymmetric Yang-Mills theory is treated in the light-cone gauge. The ensuing superfield formalism is described and is used to prove that the theory is ultra-violet finite in the perturbation expansion. The possibilities to use this field theory as a physical model is commented on, as well as the implications of the finiteness proof on N=8 supergravity.

1. Introduction

Supersymmetry was introduced into physics by Ramond[1)] in his treatment of dual models with particles of half-integer spin and by Gol'fand and Likhtman[2)] in their construction of the first supersymmetric 4-dimensional field theory. The dual model implementing this supersymmetry constructed by Neveu and Schwarz[3)] had at least three distinct advantages compared to the Veneziano model[4)]

(i) It existed in 10 dimensions of space-time rather than 26.

(ii) It allowed for half-integer spins.

(iii) It decoupled certain tachyon states (although introducing some new ones).

The dual models have an underlying interpretation in terms of relativistic strings. The supersymmetry is a symmetry of the 2-dimensional world sheet swept out by the string. In this context, it was realized for the first time that since supersymmetry demands an equal number of bosonic and fermionic states, there is a chance of cancellations among quantum corrections. In fact, for the spinning string with spin-$\frac{1}{2}$ it was found[5] that the zero-point fluctuations exactly cancel, leaving the quantum state with the same mass as the classical one, in contrast to the spin-0 state, which has non-supersymmetric boundary conditions.

The introduction of supersymmetry into field theory was also soon realized to lead to theories with improved quantum properties[6]. A key observation was made by Zumino[7], who showed that the sum of vacuum diagrams cancel completely. The supersymmetry could also be extended. This was particularly important for Yang-Mills and supergravity theories. A natural question to ask was then whether the theories with extended supersymmetry have even further improved quantum properties. In this context, it was conjectured by Gell-Mann and Schwarz[8] that the Yang-Mills theory with maximal supersymmetry (N=4) [9] is an ultraviolet finite quantum field theory. This being true would be fascinating for at least two reasons.

(i) It would be the first finite non-trivial quantum field theory in 4 dimensions of space-time.

(ii) It could indicate that the theories with maximal supersymmetry have special UV behaviour and hence that the other such theory of importance, N=8 supergravity, be also a finite quantum theory.

In the case of the N=4 model, it was soon proved that the one-loop and two-loop contributions to the charge renormalization function are zero[10]. The model contains 1 vector field, 4 spinor fields and 6 scalar fields all in the adjoint representation of the gauge group. The Lagrangian in terms of these fields is quite complicated and it is difficult to perform extensive quantum calculations in this formalism. Hence it was clear that in order to further study the model, a more efficient formalism was needed. The natural candidate for such a one is the superspace approach[11], where the supersymmetry and the supermultiplets are directly built into superfields. The corresponding such formalisms were known to exist for N=1 [12] and N=2 [13] supersymmetric Yang-Mills theories. However, in spite of great efforts by many people, no such formalism could be found for the N=4 model. In fact quite strong arguments were eventually advanced suggesting that such a formalism may not even exist for this model[14].

In the meantime, a new development occurred. The old Neveu-Schwarz-Ramond dual model was known to have inconsistencies such as tachyons. It was conjectured by Gliozzi, Olive and Scherk[15] that the model could be made consistent by projecting out some of the states. This program was implemented by Green and Schwarz[16]. They constructed a new spinning string model "the superstring", where this projection is automatic. One key element in the formalism of Green and Schwarz is the use of the light-cone gauge. In the old dual models one could either use a covariant formalism or a light-cone one. However, no covariant formalism for the superstring has yet been possible to find. This could in fact mean that the theory can only be constructed in the light-cone gauge.

It was then realized by the author, Lindgren and Nilsson[17] and also by Mandelstam[18] that the light-cone gauge could also be a way out of the dilemma with the N=4 Yang-Mills theory. By choosing this gauge, one can eliminate all unphysical degrees of freedom and it can be seen that the dynamical degrees of freedom do represent a supermultiplet of the full N=4 supersymmetry. This leads to the description of the theory in terms of an index-free superfield and the Lagrangian in terms of this superfield becomes quite manageable. The UV finiteness follows from a detailed power-counting argument for a general Green's function. Here we will describe this method and also indicate a general formalism for constructing any supersymmetric field thoery in the light-cone frame. From this we will conclude that the finiteness of the N=4 theory does not imply finiteness for the N=8 supergravity theory. However, it will seem probable that the formalism suitable to investigate the quantum properties of this model is in fact the light-cone gauge one.

There exists now also an alternative proof of the finiteness using covariant methods[19]. This is based on the use of N=2 superfields and is described in the article by P.S. Howe in this volume.

2. N=1 Supersymmetric Yang-Mills Theory in the Light-Cone Gauge.

In order to show how a covariant and a light-cone superspace formalism work, I shall describe the simplest supersymmetric Yang-Mills theory. It describes one vector field and one spinor field both in the adjoint representation of the gauge group. Its action is

$$S=\int d^4x\left[-\frac{1}{4}F_{\mu\nu}^{a}F^{\mu\nu a}+\frac{i}{2}\bar{\lambda}^a\gamma^\mu D_\mu\lambda^a+\frac{1}{2}D^aD^a\right]. \tag{2.1}$$

(For notations and conventions, see Appendix.)

The field D^a is a non-dynamical field, whose sole purpose in the action is to make it supersymmetric, i.e. the fields $A_\mu{}^a$, λ^a and D^a span a supermultiplet. Knowing the full supermultiplet, it is not hard to find a covariant superfield encompassing the whole supermultiplet[12]. In this case, one uses a real superfield with a guage invariance under transformations with a chiral superfield. I will not describe this formalism further since it has been described extensively in the literature. See also the article by P.S. Howe.

This superfield formalism is quite adequate to use to describe the theory, and it has been used to derive Feynman supergraphs, etc. When we turn to the N=4 model, we know the dynamical fields (and their couplings) and there has been an extensive search for a set of auxiliary fields which would complete the supermultiplet off-shell. One could either try directly to find this set and then construct the corresponding superfield or one could search directly for the superfield. However, no such attempt has been successful. In fact, there are strong arguments[14] that no such superfield exists.

There is certainly no guarantee that a superfield has to exist. We must now remember that not all the components of $A_\mu{}^a$ and λ^a are dynamical. There are also extra auxiliary components which are introduced only to make the Lorentz symmetry covariant. What we should ask is what happens if we eliminate those, too? The way to implement this scheme is to use the light-cone gauge. We choose

$$A^{+a} = 0. \tag{2.2}$$

With this gauge choice, it is not necessary to introduce Feynman-Fade'ev-Popov ghosts, since it is a physical gauge and the gauge choice can be directly implemented in the action. With this gauge choice, we have broken the explicit Lorentz covariance and we can only expect to have SO(2) covariance left. Hence we split the spinor into its two "light-cone" SO(2) spinors

$$\lambda = \frac{1}{2}(\gamma_+\gamma_- + \gamma_-\gamma_+)\lambda \equiv \lambda_+ + \lambda_- . \tag{2.3}$$

Another ingredient is the use of the light-cone (or null-plane) frame[20]. This means that we regard

$$x^+ = \frac{1}{\sqrt{2}}(t + z) \tag{2.4}$$

as the evolution ("time") variable and

$$x^- = \frac{1}{\sqrt{2}}(t - z) \tag{2.5}$$

as a "space" variable.

With the introduction of (2.2) and (2.3) into the action, one finds that $A^- = -A_+$ and λ_- are non-propagating, i.e. their equations of motion do not involve $\partial/\partial x_+ = \partial_+ = -\partial^-$. In this way, we have found the auxiliary components of the vector and spinor fields. They occur in quadratic and linear terms and to simplify matters, we complete the squares by introducing

$$S^a = \partial^+ A_+^a + \partial^i A^{ia} + g f^{abc}\left[\frac{1}{\partial^+}(A_i^b \partial^+ A_i^c) + \frac{i}{2}\frac{1}{\partial^+}(\bar{\lambda}_+^b \gamma^+ \lambda_+^c)\right], \tag{2.6}$$

$$\eta_-^a = \lambda_-^a + \frac{1}{2\partial^+}\gamma_- \gamma^\mu \partial_\mu \lambda_+^a - \frac{1}{2} g f^{abc} \frac{1}{\partial^+}(\gamma_- \gamma^i \lambda_+^c A^{ib}), \tag{2.7}$$

where we have also introduced the non-local operator $(\partial^+)^{-1}$ defined by

$$\frac{1}{\partial^+} f(x^-) = \frac{1}{2}\int d\xi\, \epsilon(\xi - x^-) f(\xi). \tag{2.8}$$

Finally, we use a U(1)-invariant formulation. This is achieved by introducing

$$A^a \equiv \frac{1}{\sqrt{2}}(A^{1a} + i A^{2a}), \tag{2.9}$$

$$\partial \equiv \frac{1}{\sqrt{2}}(\partial_1 + i\partial_2), \tag{2.10}$$

and by choosing a representation of the spinors such that λ_+^a and η_-^a are just complex Grassmann fields χ^a and ϕ^a. The full Lagrangian is then

$$
\begin{aligned}
S = \int d^4x \Big\{ & \bar{A}^a \Box A^a + \frac{i}{\sqrt{2}} \bar{\chi}^a \frac{\Box}{\partial^+} \chi^a \\
& + g f^{abc} \Big[-2 \frac{\bar{\partial}}{\partial^+} A^a \partial^+ \bar{A}^b A^c + i\sqrt{2} \frac{\bar{\partial}}{\partial^+} A^a \bar{\chi}^b \chi^c \\
& \qquad - i\sqrt{2} A^a \chi^b \frac{\bar{\partial}}{\partial^+} \bar{\chi}^c + \text{c.c.} \Big] \\
& + g^2 f^{abc} f^{ade} \Big[-2 \frac{1}{\partial^+} (\partial^+ A^b \bar{A}^c) \frac{1}{\partial^+} (\partial^+ \bar{A}^d A^e) \\
& \qquad - i\sqrt{2} \frac{1}{\partial^+} (\bar{\chi}^b \bar{A}^c) A^d \chi^e \\
& \qquad + i\sqrt{2} \frac{1}{\partial^+} (\partial^+ A^b \bar{A}^c + \partial^+ \bar{A}^b A^c) \frac{1}{\partial^+} (\bar{\chi}^d \chi^e) \\
& \qquad + \frac{1}{\partial^+} (\bar{\chi}^b \chi^c) \frac{1}{\partial^+} (\bar{\chi}^d \chi^e) \Big] \\
& + \frac{1}{2} S^a S^a + i\sqrt{2}\, \bar{\varphi}^a \partial^+ \varphi^a + \frac{1}{2} D^a D^a \Big\} .
\end{aligned}
\tag{2.11}
$$

It is obvious that the fields S^a, ϕ^a and D^a all play the same role as auxiliary fields necessary to implement the super-Poincaré group covariantly (and hence linearly). It is also clear that elimination of say one of the auxiliary fields will break the symmetry. However, we still retain the full super-Poincaré invariance, if we eliminate all auxiliary fields. This is easy to accomplish since in a functional integral, we just have to perform Gaussian integrals.

After elimination of the auxiliary fields, the Lagrangian is written just in terms of the dynamical fields A^a and χ^a. There are essentially no indices left. The explicit relations between the various terms are completely determined by the super-Poincaré invariance. The Lorentz transformations and the supersymmetry transformations are now non-linear and can be obtained in the following way: Start with the linear transformations of A^{ia} and λ_+^a. Perform also a gauge transformation such that the gauge choice (2.6) is still satisfied. Then substitute for A^{-a} and λ_-^a and rewrite the formulae in terms of A^a and χ^a. The non-linear Lorentz transformations are connected to the generators J^{+-} and J^{i-}. The supersymmetry generator Q we split up in its two light-cone spinors

$$Q^\alpha = Q_+^\alpha + Q_-^\alpha . \tag{2.12}$$

In our specific representation, both are complex index-free operators and we use the notation

$$\begin{aligned} Q_+ &\longrightarrow q_+ , \\ Q_- &\longrightarrow q_- . \end{aligned} \tag{2.13}$$

They satisfy the algebra

$$\begin{aligned} \{q_+ , \bar{q}_+\} &= -\sqrt{2}\, p^+ , \\ \{q_- , \bar{q}_-\} &= -\sqrt{2}\, p^- , \\ \{q_+ , \bar{q}_-\} &= -\sqrt{2}\, p . \end{aligned} \tag{2.14}$$

The correct interpretation is to use $p^- = H$, the light-cone Hamiltonian. This is natural to use since we have used equations of motion. This also means that since H is non-linear so is q_-, while q_+ still is a linearly realized generator. The explicit transformations are

$$
\begin{aligned}
\delta_+ A^a &= i\alpha \bar{\chi}^a , \\
\delta_+ \chi^a &= \sqrt{2}\,\alpha\, \partial^+ \bar{A}^a ,
\end{aligned}
\tag{2.15}
$$

$$
\delta_- A^a = i\beta \frac{\partial}{\partial^+} \bar{\chi}^a + i\beta g f^{abc} A^b \frac{1}{\partial^+} \bar{\chi}^c + i\bar{\beta} g f^{abc} \frac{1}{\partial^+}\left(\partial^+ A^b \frac{1}{\partial^+} \bar{\chi}^c\right),
$$

$$
\begin{aligned}
\delta_- \chi^a = {} & \sqrt{2}\,\beta\, \partial \bar{A}^a + \beta g f^{abc} \frac{1}{\partial^+}\left[\sqrt{2} A^b \partial^+ \bar{A}^c + i \partial^+ \chi^b \frac{1}{\partial^+} \bar{\chi}^c\right] \\
& - i\bar{\beta} g f^{abc} \chi^b \frac{1}{\partial^+} \chi^c .
\end{aligned}
\tag{2.16}
$$

The great virtue of this formalism is that q_+ is linearly realized. Hence, it is natural to build in this symmetry into a superfield. Furthermore, the dynamical fields span a supermultiplet w.r.t. this symmetry, so such a superfield encompasses <u>all</u> the fields in the action. Finally, since all fields are index-free, we expect the superfield also to be index-free.

To represent the q_+ algebra, we introduce a complex Grassmann parameter θ. Then we can write

$$
\begin{aligned}
q_+ &= -\frac{\partial}{\partial \bar{\theta}} + \frac{i}{\sqrt{2}} \theta \partial^+ , \\
\bar{q}_+ &= \frac{\partial}{\partial \theta} - \frac{i}{\sqrt{2}} \bar{\theta} \partial^+ .
\end{aligned}
\tag{2.17}
$$

A general superfield will now be a function of θ and $\bar{\theta}$. However, such a superfield will not be an irreducible representation of the q_+-algebra. In fact, we can also construct covariant derivatives d and $\bar{d}$

$$
\begin{aligned}
d &= -\frac{\partial}{\partial \bar{\theta}} - \frac{i}{\sqrt{2}} \theta \partial^+ , \\
\bar{d} &= \frac{\partial}{\partial \theta} + \frac{i}{\sqrt{2}} \bar{\theta} \partial^+ .
\end{aligned}
\tag{2.18}
$$

To find an irreducible representation, we impose a "chirality" condition

$$d\phi = 0. \tag{2.19}$$

An index-free superfield can now be written as

$$\phi^a(x,\theta,\bar\theta) = A^a(y) + i\theta\bar\chi(y), \tag{2.20}$$

with $y = (x, \bar x, x^+, x^- - i/\sqrt{2}\,\theta\bar\theta)$.

An action written in terms of ϕ^a satisfies, of course, q_+ invariance. We must, however, also impose q_- invariance. From (2.16) we derive that

$$\begin{aligned}\delta_-\phi^a = {} & \left(\beta\frac{\partial}{\partial^+}\bar q_+ - \bar\beta\frac{\bar\partial}{\partial^+}q_+\right)\phi^a + \beta g f^{abc}\frac{1}{\partial^+}\{\partial^+\phi^b\bar q_+\phi^c\} \\ & - \bar\beta g f^{abc}\frac{1}{\partial^{+2}}\{\partial^{+2}\phi^b d\bar\phi^c\}.\end{aligned} \tag{2.21}$$

We can find the Hamiltonian by commuting two δ_--transformations and then elevate this to an action. An alternative but more heuristic way is to compare directly to the action (2.11). Whatever way we use we will find the following action

$$\begin{aligned}S = \int d^4x\, d\theta d\bar\theta \Big\{ & \frac{i}{\sqrt{2}}\partial^+\bar\phi\,\Box\phi + \left(i\sqrt{2}gf^{abc}\bar\partial\phi^a\partial^+\bar\phi^b\partial^+\phi^c + c.c\right) \\ & - g^2 f^{abc}f^{ade}\frac{1}{\partial^+}(\partial^+\phi^b\partial^+ d\bar\phi^c)\frac{1}{\partial^+}(\partial^+\bar d\phi^d\partial^+\bar\phi^e)\Big\}.\end{aligned} \tag{2.22}$$

This is the light-cone action for N=1 Yang-Mills theory. From it one can derive Feynman supergraphs and compute Green's functions in a perturbation expansion.

3. The N=4 Yang-Mills Theory in the Light-Cone Gauge.

This field theory was first constructed by dimensionally reducing the 10-dimensional supersymmetric Yang-Mills theory[9]. In 4 dimensions it is described by one vector field, 4 Majorana spinors and 6 scalars all in the adjoint representation of the gauge group. That construction lacked the auxiliary fields of the supersymmetry (generalizing the D^a-field in (2.1)) and it was argued before that such fields do not exist. The natural thing in this case is then to consider the light-cone gauge action. That can be derived from the original action by integrating out the unphysical degrees of freedom in the vector and spinor fields. This was done in Ref. 17 and the following action was obtained

$$
\begin{aligned}
S = \int d^4x \Big\{ & \bar{A}^a \Box A^a + \tfrac{1}{4} \bar{C}^a_{mn} \Box C^{mna} + \tfrac{i}{\sqrt{2}} \bar{\chi}^a_m \frac{\Box}{\partial^+} \chi^{ma} \\
& + g f^{abc} \Big[-2 \frac{\bar{\partial}}{\partial^+} A^a \partial^+ \bar{A}^b A^c - \frac{1}{2} \frac{\bar{\partial}}{\partial^+} A^a \partial^+ \bar{C}^b_{mn} C^{mnc} \\
& \qquad + \tfrac{1}{2} A^a \bar{\partial} \bar{C}^b_{mn} C^{mnc} + i\sqrt{2} \frac{\bar{\partial}}{\partial^+} A^a \bar{\chi}^b_m \chi^{mc} \\
& \qquad - i\sqrt{2} A^a \chi^{mb} \frac{\bar{\partial}}{\partial^+} \bar{\chi}^c_m + i\sqrt{2} \frac{\bar{\partial}}{\partial^+} \bar{\chi}^a_m \bar{\chi}^b_n C^{mnc} + \text{c.c.} \Big] \\
& + g^2 f^{abc} f^{ade} \Big[-2 \frac{1}{\partial^+} (\partial^+ A^b \bar{A}^c) \frac{1}{\partial^+} (\partial^+ \bar{A}^d A^e) - \tfrac{1}{2} C^{mnb} A^c \bar{C}^d_{mn} \bar{A}^e \\
& \qquad - \frac{1}{2} \frac{1}{\partial^+} (\partial^+ \bar{A}^b A^c + \partial^+ A^b \bar{A}^c) \frac{1}{\partial^+} (\partial^+ \bar{C}^d_{mn} C^{mne}) \\
& \qquad - \frac{1}{16} C^{mnb} C^{pqc} \bar{C}^d_{mn} \bar{C}^e_{pq} - \frac{1}{8} \frac{1}{\partial^+} (\partial^+ \bar{C}^b_{mn} C^{mnc}) \frac{1}{\partial^+} (\partial^+ \bar{C}^d_{pq} C^{pqe}) \\
& \qquad - i\sqrt{2} \frac{1}{\partial^+} (\bar{\chi}^b_m \bar{A}^c) A^d \chi^{me} + i\sqrt{2} \frac{1}{\partial^+} (\chi^{mb} A^c) \bar{C}^d_{mn} \chi^{ne} \\
& \qquad + i\sqrt{2} \frac{1}{\partial^+} (\bar{\chi}^b_m \bar{A}^c) C^{mnd} \bar{\chi}^e_n + i\sqrt{2} \frac{1}{\partial^+} (\bar{\chi}^b_m C^{mnc}) \bar{C}^d_{np} \chi^{pe} \\
& \qquad + i\sqrt{2} \frac{1}{\partial^+} (\partial^+ A^b \bar{A}^c + \partial^+ \bar{A}^b A^c + \tfrac{1}{2} \partial^+ \bar{C}^b_{mn} C^{mnc}) \frac{1}{\partial^+} (\bar{\chi}^d_p \chi^{ep}) \\
& \qquad + \frac{1}{\partial^+} (\bar{\chi}^b_m \chi^{mc}) \frac{1}{\partial^+} (\bar{\chi}^d_n \chi^{ne}) \Big\} .
\end{aligned}
\tag{3.1}
$$

with A^a as the spin-1 field, χ^{ma} the 4 spin-½ fields and $C^{mna} = = \frac{1}{2}\varepsilon^{mnpq}\bar{C}_{pq}{}^{a}$ as the 6-spin-0 fields, transforming as $\underline{1}$, $\underline{4}$ and $\underline{6}$ under SU(4), respectively.

The action is indeed invariant under the extended super-Poincaré algebra. The supersymmetry generator $Q^{\alpha m}$ is now split into $Q^{\alpha m} = Q_+^{\alpha m} + Q_-^{\alpha m}$ and we use a representation where

$$Q_+^{\alpha m} \rightarrow q_+^m , \tag{3.2}$$

$$Q_-^{\alpha m} \rightarrow q_-^m , \tag{3.3}$$

complex Grassmann generators with no spinor indices. The explicit q_+-transformations are

$$\begin{aligned} &\delta_+ A^a = i\alpha^m \bar{\chi}_m{}^a , \\ &\delta_+ C^{mna} = -i(\alpha^m \chi^{na} - \alpha^n \chi^{ma} + \epsilon^{mnpq} \bar{\alpha}_p \bar{\chi}_q{}^a) \\ &\delta_+ \chi^{ma} = \sqrt{2}\alpha^m \partial^+ \bar{A}^a + \sqrt{2}\bar{\alpha}_n \partial^+ C^{mna} . \end{aligned} \tag{3.4}$$

We can follow exactly the route taken in the example of the N=1 theory. The supersymmetry algebra is

$$\begin{aligned} &\{q_+^m , \bar{q}_{+n}\} = -\sqrt{2}\,\delta_n^m P^+ , \\ &\{q_-^m , \bar{q}_{-n}\} = -\sqrt{2}\,\delta_n^m H , \\ &\{q_+^m , \bar{q}_{-n}\} = -\sqrt{2}\,\delta_n^m P . \end{aligned} \tag{3.5}$$

This can be represented by adjoining a Grassmann parameter θ^m with its complex conjugate $\bar{\theta}_m$ to the Minkowski space.

$$\begin{aligned} &q_+^m = -\frac{\partial}{\partial\bar{\theta}_m} + \frac{i}{\sqrt{2}}\theta^m \partial^+ , \\ &\bar{q}_{+n} = \frac{\partial}{\partial\theta^n} - \frac{i}{\sqrt{2}}\bar{\theta}_n \partial^+ . \end{aligned} \tag{3.6}$$

Similarly covariant derivatives can be constructed

$$d^m = -\frac{\partial}{\partial\bar\theta_m} - \frac{i}{\sqrt{2}}\theta^m\partial^+,$$
$$\bar d_n = \frac{\partial}{\partial\theta^n} + \frac{i}{\sqrt{2}}\bar\theta_n\partial^+. \tag{3.7}$$

An irreducible representation this time is obtained by imposing the constraints

$$d^m\phi = 0, \tag{3.8}$$

$$d^m d^n\bar\phi = \frac{1}{2}\epsilon^{mnpq}\bar d_p \bar d_q\phi. \tag{3.9}$$

The latter can also be written as

$$\bar\phi = \frac{1}{48}\frac{\bar d^4}{\partial^{+2}}\phi, \tag{3.10}$$

showing that this time the complex conjugated field is not an independent one.*

A superfield satisfying (3.8) and (3.9) can be written as

$$\phi^a(x,\theta,\bar\theta) = \frac{1}{\partial^+}A^a(y) + \frac{i}{\partial^+}\theta^m\bar\chi_m^a(y) + \frac{i}{\sqrt{2}}\theta^m\theta^n\bar C_{mn}^a(y) + \frac{\sqrt{2}}{6}\theta^m\theta^n\theta^p\epsilon_{mnpq}\chi^{qa}(y)$$
$$+ \frac{1}{12}\theta^m\theta^n\theta^p\theta^q\epsilon_{mnpq}\partial^+\bar A^a(y), \tag{3.11}$$

with $y = (x, \bar x, x^+, x^- - i/\sqrt{2}\,\theta^m\bar\theta_m)$, and we see that ϕ^a does contain all the dynamical degrees of freedom.

To construct the action in this superspace, we can follow the same two ways as in the previous case. We can either compare directly

---

*We use the convention $\bar d^4 = \epsilon^{mnpq}\bar d_m \bar d_n \bar d_p \bar d_q$ and similarly for the other fermionic operators and parameters.

to the component action (3.1) or we can commute two δ_--transformations to obtain the Hamiltonian. The δ_--transformations can be obtained from demanding closure of the super-Poincaré algebra

$$\delta_-\phi^a = \left(\beta^m \frac{\partial}{\partial^+}\bar{q}_{+m} - \bar{\beta}_m \frac{\bar{\partial}}{\partial^+} q_+^{\,m}\right)\phi^a - g f^{abc} 2\beta^m \frac{1}{\partial^+}\left[\bar{q}_{+m}\phi^b \partial^+ \phi^c\right] - g f^{abc} \frac{1}{4!}\bar{\beta}_m \frac{d^4}{\partial^{+3}}\left[q_+^{\,m} \bar{\phi}^b \partial^+ \bar{\phi}^c\right]. \tag{3.12}$$

The resulting action in superspace is

$$S = 72\int d^4x\, d^4\theta\, d^4\bar{\theta}\left\{-\bar{\phi}^a \frac{\Box}{\partial^{+2}}\phi^a + \frac{4}{3} g f^{abc}\left(\frac{1}{\partial^+}\bar{\phi}^a \phi^b \bar{\partial}\phi^c + \text{c.c.}\right) - g^2 f^{abc} f^{ade}\left[\frac{1}{\partial^+}(\phi^b \partial^+ \phi^c)\frac{1}{\partial^+}(\bar{\phi}^d \partial^+ \bar{\phi}^e) + \frac{1}{2}\phi^b \bar{\phi}^c \phi^d \bar{\phi}^e\right]\right\}, \tag{3.13}$$

where $d^4\theta$ is normalized so that $\int d^4\theta\theta^4 = 1$.

This is an appropriate action to utilize in the investigation of UV finiteness. It carries the whole N=4 supersymmetry and it is quite compact and easy to handle, and we can now turn directly to this question.

4. Feynman Supergraph Rules and the UV Finiteness of the N=4 Theory.

In order to obtain the Feynman rules, we need to know the functional derivatives w.r.t. a field satisfying (3.8) and (3.9). This we solve in complete analogy with the methods used in dealing with chiral superfields in the covariant formalism[21)] and get

$$\frac{\delta\phi^a(x,\theta,\bar{\theta})}{\delta\phi^b(x',\theta',\bar{\theta}')} = \frac{1}{(4!)^2} d^4 \delta^4(x-x')\delta^4(\theta-\theta')\delta^4(\bar{\theta}-\bar{\theta}')\,\delta^a_b, \tag{4.1}$$

$$\frac{\delta\bar{\phi}^a(x,\theta,\bar{\theta})}{\delta\phi^b(x',\theta',\bar{\theta}')} = \frac{12}{(4!)^4}\frac{\bar{d}^4 d^4}{\partial^{+2}}\delta^4(x-x')\delta^4(\theta-\theta')\delta^4(\bar{\theta}-\bar{\theta}')\delta^a_b \,. \tag{4.2}$$

The generating functional is constructed by the introduction of "chiral" sources satisfying (3.8) and (3.9) and can be written as ($z = (x,\theta,\bar{\theta})$)

$$W[J] = \int \mathcal{D}\phi^a \exp\left[i\int d^{12}z\left\{\mathcal{L} + \frac{1}{4}\phi^a\frac{\bar{d}^4}{\partial^{+4}}J^a\right\}\right]. \tag{4.3}$$

$W[J]$ is chosen so that

$$\frac{\delta^n W[J]}{\delta J^{a_1}(z_1)\ldots\delta J^{a_n}(z_n)} = (i)^n\int \mathcal{D}\phi^a\, \phi^{a_1}(z_1)\ldots\phi^{a_n}(z_n)\exp\left(i\int d^{12}z\,\mathcal{L}\right). \tag{4.4}$$

From this generating functional, we can compute the various n-point Green's functions and derive the Feynman rules. The momentum space Feynman rules are given in the Table. (Note that we only Fourier transform the x-variables. We could also Fourier transform the θ's. It does not simplify matters though.)

We first show that all Green's functions are finite by naive power counting. The question of whether such a procedure is allowed will be addressed later, where we argue that it is, in fact, valid. The first observation will be that if we disregard momenta on external legs in a supergraph all graphs are of superficial degree of divergence zero. The power-counting rules employed here are the ones appropriate for supergraphs[21] and it is convenient to do this analysis in the context of a computation of the superspace effective action. The δ-functions and the θ-integrals appearing in the propagator and in the vertex functions are then not to be taken into account in computing the dimensionality of these quantities. This fact is due to the property of supergraphs that they can always be reduced to a local expression in θ. In this process, one can show that each time a super-loop is contracted to a point in θ-space, a factor $d^4\bar{d}^4$ is absorbed, cancelling the dimension of d^4k giving the loop zero dimensionality. Similarly, we find from the Table that also the propagator and vertex functions have zero dimensionality. Hence, any supergraph will have a superficial degree of divergence equal to zero. We so have to check that there is a net positive power of external momenta on the external legs in order to make the graph UV finite. Note that from now on, the internal structure of a diagram is completely irrelevant and will not affect the arguments in the sequel.

We start by considering the part of the 3-point vertex function consisting of the first term and its permutations in the expression given in the Table. Fixing leg "1" to be external, we get in the limit k >> p

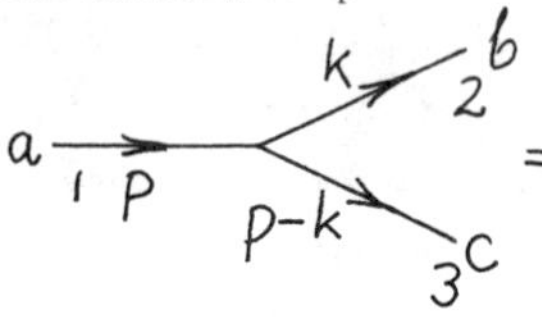

$$= \frac{i}{24} g f^{abc} \int d^8\theta \left[-\left(\frac{p-2k}{p^+} - 4\frac{k}{k^+}\right)\frac{\bar{d}_2^4\,\bar{d}_3^4}{k^{+2}k^{+2}} + \frac{k}{k^+}\frac{\bar{d}_1^4}{p^{+2}}\frac{\bar{d}_3^4}{k^{+2}} - \frac{k}{k^+}\frac{\bar{d}_1^4}{p^{+2}}\frac{\bar{d}_2^4}{k^{+2}}\right]. \quad (4.5)$$

Any supergraph having <u>all</u> external legs attached to any term in (4.5) will be finite. In the case of the first term, this follows from the fact that a partial integration of $\bar{d}_2^{\,4}$ (or $\bar{d}_3^{\,4}$) will cause it to act on the external leg (since $\bar{d}^n = 0$ for $n > 4$) removing <u>two</u> powers of momenta from the inside of the graph, giving a net power of one of external momenta. The rest of the contribution to the 3-point vertex function in (4.5) will also give finite graphs since a partial integration of $\bar{d}_2^{\,4}$ in the last term in (4.5) onto the third leg will make the worst divergent part in this term cancel the corresponding part in the second term.

Writing out the remaining terms of the 3-point vertex function in the Table, we find in the limit k >> p

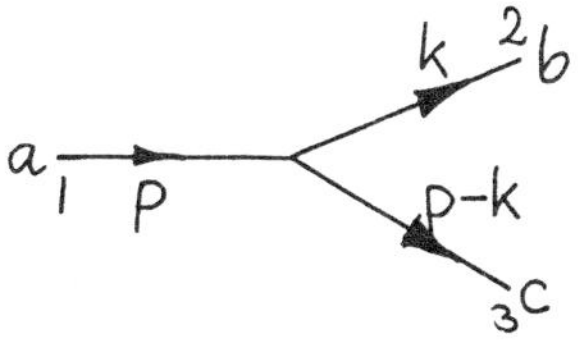

$$= -2igf^{abc}\int d^8\theta \left[-\frac{\bar{p}-2\bar{k}}{p^+}\frac{\bar{d}_1^4}{p^{+2}}+\frac{\bar{k}}{k^+}\frac{\bar{d}_2^4}{k^{+2}}-\frac{\bar{k}}{k^+}\frac{\bar{d}_3^4}{k^{+2}}\right]. \qquad (4.6)$$

The terms in (4.6) are treated in the same way as the terms in (4.5) except that the partial integration needed in order to "save" the first term in this case is performed using the d^4 coming from the internal propagators attached to the vertex. Thus all diagrams built with any combination of external vertices all coming from the 3-point vertex function are ultraviolet finite. Note that the partial integrations of $\bar{d}^4$ in the third term in (4.6) generate terms which have different $\bar{d}$'s in $\bar{d}^4$ divided between the external leg and the internal leg. These terms will, however, give rise to only UV finite diagrams.

We so have to follow the same procedure for the 4-point vertex function. Consider first the case when two legs are external. If this vertex function is written in the limit of large internal momenta one sees straightforwardly that the manipulations employed in the case of the 3-point vertex work also in this case. Thus this kind of vertex can also be used as external vertices without ever generating an infinite diagram.

The last case to consider is the 4-point vertex function with one external leg. Fixing the leg labelled 1 as external with momentum p and the internal momenta being k, ℓ and $q = -(k+\ell+p)$, we get in the limit $k,\ell >> p$:

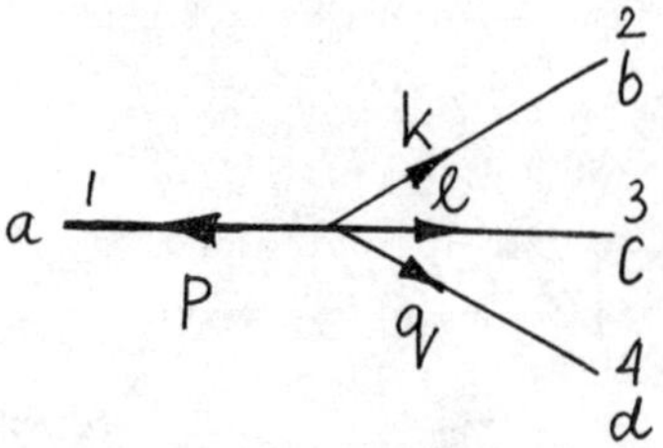

$$= -\frac{i}{32}\, g^2 \int d^8\theta$$

$$\Big\{ f^{eab} f^{ecd} \Big[\frac{q^+ - \ell^+}{q^+ + \ell^+} \Big(\frac{\bar{d}_1^4}{p^{+2}} \frac{\bar{d}_2^4}{k^{+2}} + \frac{\bar{d}_3^4}{\ell^{+2}} \frac{\bar{d}_4^4}{q^{+2}} \Big) + \frac{\bar{d}_2^4}{k^{+2}} \frac{\bar{d}_4^4}{q^{+2}} - \frac{\bar{d}_2^4}{k^{+2}} \frac{\bar{d}_3^4}{\ell^{+2}} + \frac{\bar{d}_1^4}{p^{+2}} \frac{\bar{d}_3^4}{\ell^{+2}} - \frac{\bar{d}_1^4}{p^{+2}} \frac{\bar{d}_4^4}{q^{+2}} \Big]$$

$$+ f^{eac} f^{ebd} \Big[\frac{q^+ - k^+}{q^+ - k^+} \Big(\frac{\bar{d}_1^4}{p^{+2}} \frac{\bar{d}_3^4}{\ell^{+2}} + \frac{\bar{d}_2^4}{k^{+2}} \frac{\bar{d}_4^4}{q^{+2}} \Big) + \frac{\bar{d}_1^4}{p^{+2}} \frac{\bar{d}_2^4}{k^{+2}} - \frac{\bar{d}_1^4}{p^{+2}} \frac{\bar{d}_4^4}{q^{+2}} + \frac{\bar{d}_3^4}{\ell^{+2}} \frac{\bar{d}_4^4}{q^{+2}} - \frac{\bar{d}_2^4}{k^{+2}} \frac{\bar{d}_3^4}{\ell^{+2}} \Big]$$

$$+ f^{ead} f^{ebc} \Big[\frac{\ell^+ - k^+}{\ell^+ + k^+} \Big(\frac{\bar{d}_1^4}{p^{+2}} \frac{\bar{d}_4^4}{q^{+2}} + \frac{\bar{d}_2^4}{k^{+2}} \frac{\bar{d}_3^4}{\ell^{+2}} \Big) + \frac{\bar{d}_3^4}{\ell^{+2}} \frac{\bar{d}_4^4}{q^{+2}} - \frac{\bar{d}_2^4}{k^{+2}} \frac{\bar{d}_4^4}{q^{+2}} + \frac{\bar{d}_1^4}{p^{+2}} \frac{\bar{d}_2^4}{k^{+2}} - \frac{\bar{d}_1^4}{p^{+2}} \frac{\bar{d}_3^4}{\ell^{+2}} \Big] \Big\} . \tag{4.7}$$

The terms in (4.7) divide into two distinct set of terms: those with both $\bar{d}^4$'s acting on internal legs and the rest (having one $\bar{d}^4$ acting on the external leg). The former set of terms give rise to finite graphs when used as external vertices due to the following simple calculation. Collecting the terms from, for example, the first bracket (i.e. with the common factor $f^{eab}f^{ecd}$) in (4.7) having a factor $\bar{d}_3^4\, \bar{d}_4^4$ (two of these need a partial integration in order to be cast into this form) we find

$$\bar{d}_3^4 \bar{d}_4^4 \left[\frac{q^+ - \ell^+}{q^+ + \ell^+} \frac{1}{\ell^{+2}} \frac{1}{q^{+2}} + \frac{1}{k^{+2}} \frac{1}{q^{+2}} - \frac{1}{k^{+2}} \frac{1}{\ell^{+2}} \right] = O(p), \qquad (4.8)$$

where we used momentum conservation. Note that O(p) terms lead to only finite graphs.

The final observation that establishes the finiteness is that the terms in the full vertex function that we so far have not proved to generate finite supergraphs <u>all</u> have one $\bar{d}^4$ acting on the external leg. Now, in order for a diagram to be infinite <u>all</u> its external vertices must be of this last kind and thus <u>all</u> external legs are of the same chirality. However, this implies that the diagram will not give a contribution to the superspace effective action (since its measure is over the <u>full</u> superspace) unless a factor d^4 is partially integrated onto external legs in which the diagram is finite. This completes the proof that there are no infinite Feynman supergraph in this theory in the light-cone gauge.

We have to check that we are allowed to use Weinberg's theorem, which we have used, i.e. that we can compute the degree of divergence by summing up the powers of all momenta in the integrand. This is allowed if we can Wick rotate all momenta. In fact the pole structure in p^+ following from the definition of $(\partial^+)^{-1}$ (2.8) does not permit such a rotation. However, it is clear that there is still some gauge freedom left when we have chosen the condition $A^{+a} = 0$. This freedom essentially amounts to a choice of the pole structure in p^+ [22]. In fact one can prove that one can choose a gauge in which the pole structure is $(p^+ + i\varepsilon p^-)^{-1}$. This choice is appropriate in our case since it enables us to perform the Wick rotation. Weinberg's theorem is then valid and we conclude that for this specific gauge choice, the perturbation expansion is completely finite. This implies that the charge renormalization function vanishes to all orders of perturbation theory in any gauge.

5. Extensions of the N=4 Yang-Mills Theory and Possible Applications.

Although UV finite, the model has IR divergencies. These are of exactly the same type as in an ordinary gauge theory, and supersymmetry does not improve the case[23). Another way of seeing the problem with the model is that because of the finiteness, the quantum theory is conformally invariant and hence carries no scale. This must be remedied in order to use the model as a realistic model. One way of introducing a scale is to introduce extra terms in the action[24), which break the supersymmetry but not the finiteness. In this way, one can introduce mass terms for the spinors and scalars. The way to do it is the following: Introduce a "spurion" superfield which is a constant superfield which lacks some components in its expansion in θ and carries a scale parameter. Then multiply it with a combination of the usual superfield such that mass terms are generated in the component action. One so has to check the Lorentz invariance and introduce coupling terms if necessary (since the invariance is non-linearly realized). With the new action, one can check the power counting argument again and in some cases, the model is still UV finite. It is certainly not clear that this is the only way to introduce a scale in the theory.

Having a model with this unique feature, it is really mandatory to seek ways of using it in physics. The first attempts[25) to try it as a Grand Unified Theory failed mainly due to the spinors being in the adjoint representation of the gauge group. The only group in which the adjoint and the fundamental representations are the same is E(8). However, it seems quite hard to find a symmetry breaking scheme, which results in a theory with SU(3) x SU(2) x U(1) as the gauge group with particles in the correct representations with some remnant of the finiteness left.

Another interpretation is to assume that the particles in the model are some kind of "haplons" and that quarks and leptons are bound states[26)]. This is, of course, an idea which is much more complex, and which cannot be really checked until we have reached a much more profound understanding of the non-perturbative aspects of the model. An interesting observation in this context is that the model seems to reggeize[27)], showing hints of a bound state with spin-2. Ultimately, this could be a solution also to the problem of finding a finite quantum theory for gravitation. If gravity is mediated by bound states in a finite gauge theory, there might not even be a need for a supergravity theory. I must, however, stress that so far this is only rather wild speculations and much work remains before these issues can be proved or disproved.

6. The Light-Cone Gauge Approach to N=8 Supergravity and Other Supersymmetric Theories.

In the case of the supersymmetric Yang-Mills theories, it is rather straightforward to find the light-cone superactions. Similarly we can treat the Wess-Zumino model. When we turn to supergravity, we are faced with a more difficult problem. Writing the component actions in the light-cone gauge becomes an arduous task at least for high N. The alternative way is to construct the superaction directly by demanding the super-Poincaré algebra to close and in this process, obtain the Hamiltonian.

For a gravity theory, it means that one obtains the theory expanded around a flat Minkowski metric. This does restrict the theory and should be kept in mind.

The N=8 supergravity theory has been worked out up to the 3-point coupling in the action in this way[28)]. The supermultiplet can be described by a superfield $\phi(x,\theta^m,\bar\theta_m)$ with θ^m transforming as 8 under SU(8) and hence $\bar\theta_m$ as $\bar{8}$. This superfield is constrained as in the N=4 Yang-Mills case by

$$d^m\phi = 0\,, \tag{6.1}$$

$$\phi(x,\theta,\bar\theta) = \frac{1}{8!\,2^4}\frac{\bar d^8}{\partial^{+4}}\phi(x,\theta,\bar\theta) \tag{6.2}$$

with the covariant derivatives defined as in (3.7). This field contains all the 256 degrees of freedom.

The resulting action is then found to be

$$S \sim \int d^4x\, d^8\theta\, d^8\bar\theta \Big\{ \frac{1}{4}\bar\phi \frac{\Box}{\partial^{+4}}\phi + \frac{\kappa}{3}\Big[\frac{1}{\partial^{+4}}\bar\phi \sum_{n=0}^{2}(-1)^n \binom{2}{n}(\bar\partial^{2-n}\partial^{+n}\phi\,\bar\partial^{n}\partial^{+2-n}\phi + c.c.\Big] + O(\kappa^2)\Big\}. \tag{6.3}$$

Although this is not enough to perform a conclusive power counting, it is clear that not every term in a Green's function will be finite unless new miracles occur. This is due to the dimensionality of κ. The conclusion is hence that extended supersymmetry is not enough to warrant a finite quantum gravity theory.

A remarkable possibility out of the dilemma is the superstring theory. This theory which exists only in 10 dimensions of space-time is a generalization of the N=8 theory, where the corresponding superfield ϕ is a functional of string variables $x^\mu(\sigma)$, $\theta^\alpha(\sigma)$. This model carries another dimensional parameter, the Regge slope α', and this fact gives a greater chance of obtaining finite Green's functions. The dangerous factor is the infinity of states flowing through all propagators. So far it has been proven that the one-loop contribution to the four-point S-matrix element is completely UV finite[16).

An action up to the 3-string coupling has been constructed in the same way as the supergravity theory above[29]. It is certainly too early to conclude anything about the quantum properties of this model. However, the magic appearance of a finite result at the one-loop level is quite encouraging for this model.

Appendix

We use a space-like metric for which

$$A^\mu B_\mu = A_i B_i + A_3 B_3 - A_0 B_0 = A_i B_i - A_+ B_- - A_- B_+ , \tag{A.1}$$

where i labels transverse directions and 3 and 0 the longitudinal and time directions, respectively. The Minkowski coordinates are written as

$$x = \frac{1}{\sqrt{2}} (x^1 + i x^2), \tag{A.2}$$

$$\bar{x} = \frac{1}{\sqrt{2}} (x^1 - i x^2), \tag{A.3}$$

$$x^{\pm} = \frac{1}{\sqrt{2}} (x^0 \pm x^3) \tag{A.4}$$

and similarly for the derivatives and momenta. We use the Dirac algebra

$$\{\gamma^\mu, \gamma^\nu\} = -2\eta^{\mu\nu}. \tag{A.5}$$

We write explicitly the gauge group indices which are small letters in the beginning of the alphabet, a, b, c, ...

For the internal SU(4) invariance, we use indices, which are small letters starting from m in the alphabet. Complex conjugation raises and lowers indices and the complex conjugate is denoted by a bar.

TABLE

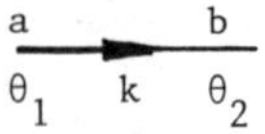

$$= \frac{2i}{(4!)^3}\,\delta^{ab}\,\frac{d_1^4}{k^2}\,\delta^8(\theta_1-\theta_2)$$

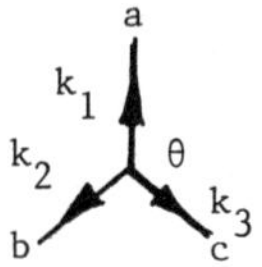

$$= 2\,igf^{abc}\int d^8\theta\ \times$$

$$\times\left\{\frac{1}{48}\,\frac{k_3}{k_1^+}\,\frac{\bar{d}_2^4}{k_2^{+2}}\,\frac{\bar{d}_3^4}{k_3^{+2}} - \frac{\bar{k}_3}{k_1^+}\,\frac{\bar{d}_1^4}{k_1^{+2}} + \text{permutations}\right\}$$

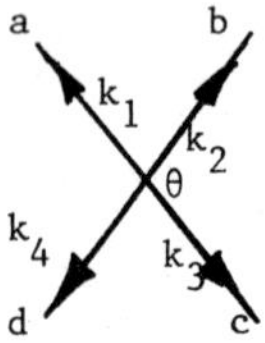

$$= -\,\frac{ig^2}{32}\int d^8\theta\ \times$$

$$\times\Bigg\{f^{eab}f^{ecd}\Bigg[\frac{k_2^+k_4^+}{(k_1^+ + k_2^+)(k_3^+ + k_4^+)}\;\frac{\bar{d}_3^4}{k_3^{+2}}\,\frac{\bar{d}_4^4}{k_4^{+2}}\ +$$

$$+\,\frac{1}{2}\,\frac{\bar{d}_2^4}{k_2^{+2}}\,\frac{\bar{d}_4^4}{k_4^{+2}}\qquad + \text{permutations}\Bigg\}$$

Explanatory comments: $\bar{d}_i^4 \equiv \bar{d}^4(k_i^+,\theta)$ where k_i is pointing away from θ and in the expression for the 3-point function $k \equiv 1/\sqrt{2}\,(k_1 + ik_2)$.

References

1. P. Ramond, Phys. Rev. D3 (1971) 2415.
2. Yu. A. Gol'fand and E.P. Likhtman, JETP Letters 13 (1971) 323.
3. A. Neveu and J.H. Schwarz, Nucl. Phys. B31 (1971) 86.
4. G. Veneziano, Nuovo Cim. 57A (1968) 190.
5. L. Brink, A Study of Dual Models in the Theory of Strong Interactions, p. 82, Thesis, Chalmers University of Technology (1973).
6. J. Wess and B. Zumino, Phys. Letters 49B (1974) 52; S. Ferrara, J. Iliopoulos and B. Zumino, Nucl. Phys. B77 (1974) 413.
7. B. Zumino, Nucl. Phys. B89 (1975) 535.
8. M. Gell-Mann and J.H. Schwarz, unpublished.
9. F. Gliozzi, J. Scherk and D. Olive, Nucl. Phys. B122 (1977) 256; L. Brink, J. Schwarz and J. Scherk, Nucl. Phys. B121 (1977) 77.
10. D.R.T. Jones, Phys. Letters 72B (1977) 199; E. Poggio and H. Pendleton, Phys. Letters 72B (1977) 200.
11. Abdus Salam and J. Strathdee, Nucl. Phys. B76 (1974) 477.
12. J. Wess and B. Zumino, Nucl. Phys. B78 (1974) 1.
13. S. Ferrara and B. Zumino, Nucl. Phys. B79 (1974) 413; P. Fayet, Nucl. Phys. B113 (1976) 135.
14. W. Siegel and M. Rocek, Phys. Letters 105B (1981) 275; B.E.W. Nilsson, Göteborg preprint 81-6 (1981); V.O. Rivelles and J.G. Taylor, Phys. Letters 121B (1983) 37.
15. See Ref. 9.
16. M.B. Green and J.H. Schwarz, Nucl. Phys. B181 (1981) 502, B198 (1982) 252, 441, Phys. Letters 109B (1982) 444.
17. L. Brink, O. Lindgren and B.E.W. Nilsson, Nucl. Phys. B212 (1983) 401, Phys. Letters 123B (1983) 323.
18. S. Mandelstam, Nucl. Phys. B213 (1983) 149.
19. P.S. Howe, K.S. Stelle and P.K. Townsend, in preparation.
20. P.A.M. Dirac, Rev. Mod. Phys. 26 (1949) 392.
21. M. Grisaru, W. Siegel and M. Rocek, Nucl. PHys. B159 (1979) 429; see also M. Rocek in "Superspace and Supergravity", Cambridge Univ. Press, eds. S.W. Hawking and M. Rocek (1981).

22. J.M. Cornwall, Phys. Rev. D10 (1974) 500.

23. M.B. Green, J.H. Schwarz and L. Brink, Nucl. PHys. B198 (1982) 474.

24. M.A. Namazie, Abdus Salam and J. Strathdee, Trieste prepring ICTP/82/230, November 1982;
A.J. Parkes and P. West, Nucl. Phys. (in press).

25. J.H. Schwarz, Orbis Scientiae 1978.

26. J.C. Pati and Abdus Salam, Trieste preprint ICTP/82/99, July 1982.

27. M. Grisaru and H. Schnitzer, Nucl. Phys. B204 (1983) 267.

28. A.K.H. Bengtsson, I. Bengtsson and L. Brink, Göteborg preprint 83-11 (1983).

29. M.B. Green, J.H. Schwarz and L. Brink, Nucl. Phys. B219 (1983) 437.

ULTRA VIOLET DIVERGENCE CANCELLATIONS IN SUPERSYMMETRIC FIELD THEORIES

P. S. Howe

Department of Mathematics
King's College London

ABSTRACT

The extended superfield approach to the quantization of supersymmetric field theories is outlined. The implications of the resulting nonrenormalization theorems for various supersymmetric field theories are discussed.

Introduction

One of the most striking features of supersymmetric field theories is that they are less ultra-violet divergent than their non-supersymmetric counterparts. This feature is present even in the simplest four-dimensional supersymmetric field theory, the Wess-Zumino model, where there is only one independent renormalization constant instead of the three one would naively expect[1)] (wave function, coupling constant and mass). It is natural to conjecture that models with extended supersymmetry should exhibit even more ultra-violet divergence cancellations and indeed this turns out to be the case, at least for renormalizable theories. It is now known that the maximally supersymmetric N=4 Yang-Mills model is not only finite[2,3,4,5)] but is also a member of an entire class of finite N=2 Yang-Mills models[6)]. Indeed, a general N=2 Yang-Mills coupled to N=2 supermatter model has only a one-loop coupling constant renormalization in background field gauges[6,7,8)].

In this talk I shall describe a general framework which is suitable for the discussion of ultra-violet divergence cancellations in any supersymmetric field theory. This approach is based on a version of the background field method adapted to superspace in which the background superfields are constrained while the quantum superfields are unconstrained[7]. Provided that certain technical conditions are fulfilled, it can be shown that the allowed counterterms in a given theory have to be expressed as integrals over the whole of superspace of gauge invariant functions of the constrained background fields. Since the dimension of the N-extended superspace measure is 2N-4 while the dimensions of the background fields are positive semi-definite, it follows that for sufficiently large N, there will be no allowed counterterms. As will be explained in the following, this argument does not apply to one-loop counterterms so that the one-loop cancellations have to be checked separately. Furthermore, there can be allowed counterterms in nonrenormalizable theories since the presence of a dimensional coupling constant invalidates the line of reasoning given above. Hence this method applied to supergravity theories merely places a lower bound on the loop order for which there are corresponding allowed counterterms.

One of the technical prerequisites for a given theory to be formulated in terms of background extended superfields is that it admit a set of auxiliary fields. A counting argument has been given which shows that in general it is not possible to find auxiliary fields for a given model with N-extended supersymmetry[9]. This means that not all of the N supersymmetries can be simultaneously linearly realized, but there will be a subset of M of them which can. Thus in general it is only possible to use M-extended superfields to describe a theory with N-extended supersymmetry where $M \leq N$. The theories we shall consider are Yang-Mills (YM) models with $N \leq 4$, $M \leq 2$, conformal supergravities (CSG) with $N \leq 4$, $M \leq 4$ and Poincaré supergravities (PSG) with $N \leq 8$, $M \leq 4$.

Extended superfield perturbation theory is technically rather complicated for several reasons. Firstly it is not possible to use constrained superfields as the quantum fields in contrast to the N=1 case where the use of chiral superfields simplifies the superspace Feynman rules considerably. The fact that one is forced to use unconstrained superfields in turn implies that there are a large number of gauge invariances even for supersymmetric matter (spins $\leq \frac{1}{2}$). Secondly, the constraints on the superspace field strengths for Yang-Mills and supergravity are difficult to solve in closed form, and it has been necessary to develop an iterative technique[8,10]. Thirdly, the large number of gauge invariances requires the introduction of a correspondingly large number of ghosts, including special ghosts which are needed only in the background field method. For these reasons, we shall present only an outline of the central ideas and illustrate them with N=1 examples. The interested reader will find a detailed account of the technicalities involved in refs. 4 and 8.

The Background Field Method in Superspace

In an ordinary gauge theory, the background field method[11] involves splitting the gauge field A_μ into background and quantum parts,

$$A_\mu = A_\mu^B + A_\mu^Q$$

$$A_\mu = A_\mu^r T^r ; \quad T^r = (T^r)^\dagger ; \quad [T^r, T^s] = i f^{rst} T^t \tag{1}$$

and then integrating only over the quantum field in the functional integral. The generating functional of connected Green's functions is given by

$$e^{iW[A_\mu^B, J_\mu]} = \int dA_\mu^Q \exp i\left\{S[A_\mu^B + A_\mu^Q] - \frac{1}{2\alpha}G^2 + J_\mu A^{\mu Q} + \text{ghosts}\right\} \tag{2}$$

and is invariant under background gauge transformations

$$\delta A_\mu^B = -\mathcal{D}_\mu^B K^B \equiv -(\partial_\mu K^B + ig[A_\mu^B, K^B]),$$
$$\delta A_\mu^Q = ig[K^B, A_\mu^Q],$$

if the gauge-fixing function G is chosen appropriately, e.g., $G = \mathcal{D}_\mu^B A^{\mu Q}$. The Legendre transform of W is given by

$$\Gamma[\tilde{A}^Q_\mu, A^B_\mu] = W[A^B_\mu, J_\mu] - J_\mu \tilde{A}^{\mu Q},$$
$$\tilde{A}^Q_\mu = \frac{\delta W}{\delta J^\mu}. \tag{3}$$

All the information one requires may be extracted from the background effective action $\Gamma\{0, A^B_\mu\}$ which is obtained by evaluating all I P I graphs with no external quantum lines. The Feynman rules are a little more complicated than usual because there are vertices involving the background fields as well as just the quantum fields, but the advantage of the method is that manifest background gauge invariance may be maintained at the quantum level given an invariant regularization scheme. It is not necessary to renormalize the quantum fields (including the ghosts) since the corresponding renormalization constants cancel between propagators and vertices so that there is only a wave-function renormalization for the background gauge field (Z_A) and a coupling constant renormalization (Z_g). Manifest background gauge invariance implies that the only counter term needed is $S\{A^B_\mu\}$ so that

$$Z_g = Z_A^{-1/2}. \tag{4}$$

The standard superspace formulation of N=1 supersymmetric Yang-Mills is in terms of a real superfield V with the gauge transformation[12)]

$$e^{gV} \longrightarrow e^{ig\bar{\Lambda}} e^{gV} e^{-ig\Lambda} \tag{5}$$

where the parameter Λ is chiral,

$$D_\alpha \bar{\Lambda} = 0, \quad D_\alpha = \frac{\partial}{\partial\theta^\alpha} - \frac{i}{2}(\sigma^\mu)_{\alpha\dot{\alpha}} \bar{\theta}^{\dot{\alpha}} \partial_\mu . \tag{6}$$

It is possible to discuss the background/quantum splitting directly in terms of unconstrained superfields[13)], but it is more convenient to use the differential geometric superspace formulation of YM[14] (in fact this is essential for $N > 1$). We therefore introduce a superspace gauge potential $A_A = (A_\mu, A_\alpha, \bar{A}_{\dot{\alpha}})$ which has the gauge transform

$$\delta A_A = -\mathcal{D}_A K \equiv -(D_A K + ig[A_A, K])$$
$$D_A = (\partial_\mu, D_\alpha, \bar{D}_{\dot{\alpha}}); \quad A_A = A_A^r T^r. \tag{7}$$

The superspace field strength F_{AB} is given by

$$F_{AB} = D_A A_B - D_B A_A (-1)^{AB} + t_{AB}{}^C A_C + ig[A_A, A_B\} \tag{8}$$

where

$$t_{AB}{}^C = \begin{cases} t_{\alpha\dot{\beta}}{}^\mu = -i(\sigma^\mu)_{\alpha\dot{\beta}} \\ \text{zero otherwise} \end{cases} \tag{9}$$

so that the (graded) commutator of two supercovariant derivatives is given by

$$[\mathcal{D}_A, \mathcal{D}_B\} = ig F_{AB} - t_{AB}{}^C \mathcal{D}_C. \tag{10}$$

Since the superfields A_μ and A_α are unconstrained and carry additional Lorentz indices, it follows that they define reducible representations of supersymmetry and constraints must be imposed on the F_{AB}'s. These constraints are

$$F_{\alpha\dot{\beta}} = 0 \quad \Rightarrow \quad A_\mu = A_\mu(A_\alpha, \bar{A}_{\dot{\alpha}}) \tag{11}$$

and

$$F_{\alpha\beta} = 0. \tag{12}$$

In N=1 these constraints are easy to solve since (12) implies that it is (complex) pure gauge

$$A_\alpha = -\frac{i}{g} e^{igU} D_\alpha e^{-igU}. \tag{13}$$

However, in extended supersymmetry, the corresponding constraints are difficult to solve in closed form. To illustrate how they can be solved iteratively[8,10], let us differentiate (8) with respect to the coupling constant g,

$$\frac{\partial}{\partial g}(gF_{AB}) = \mathcal{D}_A B_B - (-1)^{AB}\mathcal{D}_B B_A + t_{AB}{}^C B_C ; \quad B_A = \frac{\partial}{\partial g}(gA_A). \tag{14}$$

This manoeuvre removes the commutator term from (8) thereby making it easier to solve. From (12) and (14),

$$\frac{\partial}{\partial g}(gF_{\alpha\beta}) = 0 \Rightarrow \mathcal{D}_\alpha B_\beta + \mathcal{D}_\beta B_\alpha = 0$$
$$\Rightarrow B_\alpha = -\mathcal{D}_\alpha U \tag{15}$$

and

$$A_\alpha = \frac{1}{g}\int_0^g dk\, B_\alpha(k). \tag{16}$$

A_α may be constructed order by order in g from equation (16). In N=1 it is straightforward to do the integration in (16) to get (13) but in general it will not be possible to give a closed form expression for the constrained potentials. Notice that in solving the constraint (12), a new gauge transformation has been introduced:

$$e^{igU} \rightarrow e^{igK} e^{igU} e^{-ig\bar{\Lambda}}, \quad D_\alpha \bar{\Lambda} = 0. \tag{17}$$

It is the "pregauge" transformation Λ which is important at the quantum level since the original K gauge transformation can be used to set the real part of U to zero. After such a gauge choice, one has

$$e^{igU} = e^{-gV/2} \tag{18}$$

so that the usual N=1 YM superfield V is recovered. The background-quantum split can be performed at the level of the potentials by putting[7)]

$$A_\alpha = A_\alpha^B + A_\alpha^Q . \tag{19}$$

Then we have

$$F_{\alpha\beta} = F_{\alpha\beta}^B = 0 \tag{20}$$

$$\Rightarrow \mathcal{D}_\alpha^B A_\beta^Q + \mathcal{D}_\beta^B A_\alpha^Q + ig\{A_\alpha^Q, A_\beta^B\} = 0. \tag{21}$$

(21) can be solved for A_α^Q:

$$A_\alpha^Q = -\frac{i}{g} e^{igU^Q} \mathcal{D}_\alpha^B e^{-igU^Q} , \tag{22}$$

where U^Q is the unconstrained quantum prepotential. (Again (quantum) K-gauge invariance can be used to set $U^Q = i(V^Q/2)$ if desired.).

The N=1 YM action is given by

$$I = \mathrm{tr}\int d^4x\, d^2\theta\, W^\alpha W_\alpha + c.c. \tag{23}$$

where

$$(\sigma^\mu)_\alpha{}^{\dot\alpha} F_{\dot\alpha\mu} = 2W_\alpha \equiv 2(W_\alpha^B + W_\alpha^Q) . \tag{24}$$

Although the action (23) is a chiral integral, it is in fact possible to rewrite it as a full superspace integral without using the explicit form (13) for A_α [8].

In order to formulate extended YM and supergravity theories in terms of extended superfields, one proceeds in the same way as for N=1 YM, although it will be necessary to use the iteration method to solve the higher N analogues of the constraints (11) and (12). In addition there are supermatter multiplets (spins $\leq \frac{1}{2}$) which exist only for $N \leq 2$. The N=1 supermatter multiplet is just the chiral multiplet, while the N=2 supermatter multiplet (the hypermultiplet) has a physical spectrum which is double that of the N=1 chiral multiplet. The N=2 superfield version of this multiplet is rather complicated, but necessary for the formulation of the general N=2 YM-matter model (including N=4 YM) in terms of N=2 superfields [4]. Common to all these theories in the background field approach is the fact that the background fields are constrained (e.g. A_α^B in N=1 YM), while the quantum fields are unconstrained (e.g. V^Q in N=1 YM).

Special Features At One Loop

One loop is special in the background field superspace method because of complications due to the ghost structure of theories quantized in terms of unconstrained superfields. To illustrate this consider a chiral superfield coupled to N=1 YM

$$\bar{\mathcal{D}}_{\dot\alpha}\phi = \bar{D}_{\dot\alpha}\phi + ig\bar{A}_{\dot\alpha}\phi , \tag{25}$$

where ϕ is in a representation of the gauge group

$$\delta\phi = iK\phi \; ; \; K = K^r T^r . \tag{26}$$

We want to quantize ϕ in terms of an unconstrained prepotential; this is not strictly necessary in N=1 but illustrates the general situation for higher N. In the absence of interactions we have

$$\bar{D}_{\dot{\alpha}}\phi = 0 \Rightarrow \phi = \bar{D}^2\bar{\chi} \tag{27}$$

where χ is complex and unconstrained. Clearly ϕ is invariant under the gauge transformation

$$\delta\bar{\chi} = \bar{D}^{\dot{\alpha}}\bar{\Lambda}_{\dot{\alpha}}. \tag{28}$$

In the interacting case, ϕ can be split into background and quantum parts as follows,

$$\phi = e^{ig\bar{U}^Q}[\phi^B + \bar{\mathcal{D}}^{B2}\bar{\chi}], \tag{29}$$

where U^Q is the quantum YM prepotential. The background field ϕ^B is covariantly chiral with respect to the background covariant derivative

$$\bar{\mathcal{D}}^B_{\dot{\alpha}}\phi^B = 0$$

and χ is the quantum matter prepotential. ϕ is left invariant under the gauge transformation

$$\delta\bar{\chi} = \bar{\mathcal{D}}^{\dot{\alpha}B}\bar{\Lambda}_{\dot{\alpha}}. \tag{30}$$

The gauge invariance (30) has to be fixed in order to quantize and a suitable gauge fixing function is

$$F_\alpha = \mathcal{D}_\alpha{}^B\bar{\chi}. \tag{31}$$

Application of the usual Fadeev-Popov procedure for this gauge invariance gives rise to the following ghost action

$$I_{ghost} = \int d^4x\, d^4\theta\, C'^\alpha \mathcal{D}_\alpha{}^B \bar{\mathcal{D}}^{\dot{\alpha}B}\bar{C}_{\dot{\alpha}} + c.c. \tag{32}$$

where C_α and C'_α are commuting Fadeev-Popov ghosts. The action (32) has new gauge invariances

$$\delta C_\alpha = \mathcal{D}^{\beta\gamma}\Lambda_{\gamma\alpha}\ ; \quad \Lambda_{\gamma\alpha} = \Lambda_{(\gamma\alpha)}\ ; \tag{33a}$$

$$\delta C'_\alpha = \mathcal{D}^{\beta\gamma}\Lambda'_{\gamma\alpha}\ ; \quad \Lambda'_{\gamma\alpha} = \Lambda'_{(\gamma\alpha)}\ . \tag{33b}$$

It is clear that this process can be repeated indefinitely leading to an infinite sequence of Fadeev-Popov ghosts[15]. The first generation ghosts couple to the Yang-Mills ghosts (since χ also transforms under quantum Λ transformations), but all the others couple only to the background field A^B_α via the covariant derivatives. These couplings can only contribute to one loop graphs, so that the one loop order in background superfield perturbation theory is indeed special.

In order to define the theory at one loop, it is necessary to truncate the infinite ghost sequence. This can be done by rewriting (33a) in the form

$$\delta C_\alpha = e^{igU^B} D^\beta \tilde{\Lambda}_{\beta\alpha}\ , \tag{34}$$

where U^B is the background prepotential

$$A^B_\alpha = -\frac{i}{g}\, e^{igU^B} D_\alpha\, e^{-igU^B} . \tag{35}$$

Corresponding to the gauge invariance (34) there will be a pair of Fadeev-Popov ghosts $C_{\alpha\beta}$ and $C'_{\alpha\beta}$, but they will always appear in the forms $D^\alpha C_{\alpha\beta}$, $D^\alpha C'_{\alpha\beta}$. Hence the second generation ghost action will be invariant under gauge transformations

$$\delta C_{\alpha\beta} = D^\gamma \Lambda_{\alpha\beta\gamma}\ ; \quad \Lambda_{\alpha\beta\gamma} = \Lambda_{(\alpha\beta\gamma)} \tag{36}$$

and similarly for $\delta C'_{\alpha\beta}$. An identical construction has also to be made for the transformation (33b)

Since the transformations (36) involve no background fields, it follows that the third and higher generation ghosts decouple.

To summarize, the first generation Fadeev-Popov ghosts couple to the constrained background fields and the unconstrained quantum fields, the second generation couple to the YM background prepotential only and contribute only to one-loop graphs, and the third and higher generations decouple. In addition, the background field method requires the introduction of further Nielsen-Kallosh ghosts which couple to the background YM prepotential at the one loop level only[16)]. These arise from the 't Hooft averaging trick which is used to get the gauge fixing functions appearing quadratically in the action.

The general features of the one loop structure described here for N=1 chiral matter are in fact common to all supersymmetric field theories in the background field method. In order to truncate the infinite ghost sequence it is necessary to be able to rewrite the transformation (33) in the form (34), i.e. to be able to "convert" covariant derivatives into flat derivatives. In general, this is a complicated procedure and has to be done iteratively, as is discussed in ref. 8. The importance of this construction lies in the fact that it is necessary to introduce background prepotentials for supergravity and Yang-Mills fields. Since these background prepotentials have negative or zero dimensions, this spoils the argument given in the introduction for the absence of counterterms, but only at the one-loop level.

The Non-Renormalization Theorems.

A general N-extended four dimensional supersymmetric field theory can be formulated in terms of M-extended superfields describing supergravity, Yang-Mills and supermatter. The constrained background fields will be denoted by Φ^B (for YM Φ^B is A^i_α, i = 1, .. M,) for supergravity, it is the superspace vielbein $E_M{}^A$ and for supermatter, the scalar superfields whose $\theta = 0$ components correspond to the dimension one scalars). In order to derive the non-renormalization theorems, it is necessary that the following conditions hold: (i) that the part of the action from which the Feynman Rules are derived should be an integral over the whole of M-extended superspace and (ii) that the background fields should occur in this part of the action only via the constrained fields Φ^B, except for terms involving second generation Fadeev-Popov ghosts and Nielsen-Kallosh ghosts. If these conditions are met, then it follows that each contribution to the effective action will be of the form

$$\int d^4x_1 \ldots d^4x_n \, d^{4M}\theta \, f(x_1 \ldots x_n, \theta). \tag{37}$$

At L ≥ 2 loops, the counterterm actions must therefore have the form

$$I = \int d^4x \, d^{4M}\theta \, \mathcal{L}(\Phi^B) \tag{38}$$

where $L(\Phi^B)$ is a gauge invariant function of the background fields Φ^B. At L=1 loop the counterterms may involve the background YM and supergravity prepotentials because of the special one loop features discussed in the last section.

Equation (38) is the key equation in the superspace background field method from which the non-renormalization theorems follow by straightforward dimensional analysis. The simplest example occurs in the Wess-Zumino model where the mass and potential terms are not renormalized[1]. The background field in this case is the chiral superfield ϕ^B. Since the mass and potential terms are integrals only over chiral superspace (when written in terms of ϕ), there are no corresponding counterterms which are integrals over the whole superspace of functions of ϕ^B.

In ordinary quantization, the chiral field ϕ is related to the prepotential X by

$$\phi = \bar{D}^2 \bar{X}. \tag{39}$$

Hence in order to get a ϕ field on an external line, it is necessary to remove a $\bar{D}^2$ from the graph under consideration. Since D has dimension ½ this will improve the convergence properties of the corresponding Feynman integral. In the background field method this mechanism is built into the formalism so that UV divergence cancellations can be deduced without it being necessary to examine graphs in detail.

Equation (38) presupposes that it is possible to regulate the theory in a gauge invariant supersymmetric way. For $L \geq 2$ loops, this can be done by using higher derivative regularization, but this will have to be supplemented by a separate one loop regularization scheme.

General Results

(i) Yang-Mills-Matter

All N=2 and 4 Yang-Mills Matter models can be formulated in terms of N=2 extended superfields in such a way that the requirements of the last section are met. In N=1, the Yang-Mills action itself can be written as a full superspace integral involving A_α so there is no non-renormalization theorem. However, if the gauge group has a U(1) factor, there is the possibility that the Fayet-Iliopoulos term could be generated as a counterterm. Since

$$I_{F.I} \propto \int d^4x\, d^4\theta\, V, \tag{40}$$

it follows that this can only happen at one loop, as has been shown previously by explicit calculations[17].

In N=2 models, the background YM field is A^i_α (i = 1, 2) and has dimension ½. It is introduced in a similar way to A_α in N=1. The background matter superfields have dimension 1 (their leading components are the dimension one scalars) and there are no matter self-outeraction terms[4]. From (38) we have immediately that there are no $L \geq 2$ counterterms in any N=2 YM model. Furthermore, since the non-renormalization theorem holds at one loop in the matter sector the only independent renormalization that is necessary is a one loop gauge coupling constant renormalization. In order to get a finite N=2 model, it is therefore sufficient to arrange the matter such that the one-loop β-function vanishes. For m_i hypermultiplets in representations R_i of the gauge group we require[6)]

$$\sum m_i T(R_i) = C_2(G), \tag{41}$$

where

$$\mathrm{tr}\, T^r T^s = \delta^{rs} C_2(R) \;, \quad T^r T^r = T(R)\mathbb{1} \tag{42}$$

or a representation R (G signifies the adjoint representation). Although the currently available off-shell version of the hypermultiplet cannot be put into a complex representation of the gauge group without doubling the number of physical degrees of freedom[4)], an explicit two loop calculation suggests that if (41) holds the theory is finite even for hypermultiplets in complex representations[18)]. This suggests that there may be an alternative off-shell version of the hypermultiplet.

(ii) Conformal Supergravity

The graviton part of the conformal supergravity action is the Weyl action (Weyl tensor squared) which is fourth order in derivaties. Hence although these theories are power counting renormalizable, it is not clear that they are unitary. Conformal supergravity theories exist for $N \leq 4$ and all of them admit auxiliary fields[19)]. The corresponding superspace constraints are also known[20)] so it is almost certain that the theory can be formulated in such a way that equation (38) is applicable. Since the background supergravity fields have dimension greater or equal to zero, the only extended CSG which has any $L \geq 2$ loop allowed counterterm is N=2 where

$$I = \int d^4x\, d^8\theta\, E \tag{43}$$

where E is the superdeterminant of the vielbein matrix. However, the volume of N=2 superspace vanishes identically[21] so that there are no $L \geq 2$ loop counterterms in any extended conformal supergravity theory. The N=4 theory has been shown to be finite at one loop by explicit calculation[22] so it seems likely that this theory is finite to all orders. It may also be possible to construct finite N=2 and 3 models by adding superconformal matter in such a way as to ensure one loop finiteness.

(iii) Poincaré Supergravity

Poincaré supergravity theories are quite different from the models we have considered previously in that they are non-renormalizable. A general L-loop counterterm for a supergravity theory quantized in terms of M-extended superfields has the form

$$I^{(L)} = \kappa^{2(L-1)} \int d^4x\, d^{4M}\theta\, \mathcal{L}^{(L)} \tag{44}$$

with

$$\dim \mathcal{L}^{(L)} = 2(L - M + 1). \tag{45}$$

The general counterterm structure for N extended supergravity was analyzed in terms of on-shell superfields some time ago, and it was shown that there are counterterms of the form (44)[23,24] with N=M starting at (n - 1) loops for $N \geq 5$ and 3 loops for $N \leq 4$[23,24]. However, as has been emphasized, it is not possible to quantize N extended supergravity in terms of N extended superfields but rather one must use M extended superfields where N and M are related by the following table

N	8	6	5	4	3	2	1
M	4	3	3	2	2	2	1

(46)

When this fact is taken into account, it turns out that there are allowed counterterms at three loops for all supergravity theories. It is in fact not difficult to show that the linearized three loop N=8 counterterm[24,25)] can be written as an integral over the whole of M=4 superspace of an expression involving M=4 superfields[8)].

In conclusion, the background field method in superspace is a powerful tool for investigating the ultra violet divergence cancellations of supersymmetric field theories. There is a class of finite N=2 Yang-Mills models, including N=4 as a special case, and at least one finite conformal supergravity theory (N=4). On the other hand, the method allows counterterms starting at the three loop order for all Poincaré supergravity theories. It is possible that the co-efficients of these counterterms could turn out to be zero in an actual calculation, but such cancellations would have to involve an entirely new mechanism form the one responsible for those currently known.

Acknowledgement

This talk is based on work done in collaboration with K.S. Stelle and P.K. Townsend (refs. 4 and 8).

References

1. J. Wess and B. Zumino, Phys. Lett. 49B (1974) 52;
J. Iliopoulos and B. Zumino, Nucl. Phys. B76 (1974) 310;
S. Ferrara, J. Iliopoulos and B. Zumino, Nucl. Phys. B77 (1974) 413.
2. D.R.T. Jones, Phys. Lett. 72B (1977) 199;
E. Poggio and H. Pendleton, Phys. Lett. 72B (1977) 200;
O.V. Tarasov and A.A. Vladimirov, Phys. Lett. 96B (1980) 94;
M.T. Grisaru, M. Rocek and W. Siegel, Phys. Rev. Lett. 45 (1980) 94;
W. Caswell and D. Zanon, Phys. Lett. 100B (1980) 152.
3. S. Ferrara and B. Zumino, unpublished;

M.F. Sohnius and P.C. West, Phys. Lett. 100B (1981) 245;
K. Stelle in "Quantum Structure of Space and Time" eds. M.J. Duff and C.J. Isham (C.U.P. 1982) 337.

4. P.S. Howe, K.S. Stelle and P.K. Townsend, Nucl. Phys. B214 (1983) 519.
5. S. Mandelstam, Nucl. Phys. B213 (1983) 149.
   L. Brink, O. Lindgren and B. Nilsson, University of Texas preprint UTTG-1-82.
6. P.S. Howe, K.S. Stelle and P.C. West, Phys. Lett. 124B (1983) 55.
7. M.T. Grisaru and W. Siegel, Nucl. Phys. B201 (1982) 292.
8. P.S. Howe, K.S. Stelle and P.K. Townsend, Imperial College preprint ICTP/82-83/20 (1983).
9. V.O. Rivelles and J.G. Taylor, J. Phys. A15 (1982) 163.
10. J. Koller, Caltech preprint (1983).
11. G. 't Hooft in Proc. XII Winter School in Karpacz, Acta Universitalis Wratislarensis No. 38, 1975;
    B.S. De Witt in Quantum Gravity 2, eds. C.J. Isham, R. Penrose, D.W. Sciama (O.U.P. 1981) 449;
    D.G. Boulware, Phys. Rev. D23 (1981) 389;
    L. Abbot, Nucl. Phys. B185 (1981) 189.
12. A. Salam and J. Strathdee, Nucl. Phys. B86 (1975) 142;
    S. Ferrara and O. Piguet, Nucl. Phys. B93 (1975) 261.
13. M.T. Grisaru, W. Siegel and M. Rocek, Nucl. Phys. B159 (1979) 429.
14. J. Wess, Lecture Notes in Physics, 77 (Springer-Verlag 1978).
15. W. Siegel and S.J. Gates, Nucl. Phys. B189 (1981) 295.
16. N.K. Nielsen, Nucl. Phys. B140 (1978) 499;
    R.E. Kallosh, Nucl. Phys. B141 (1978) 141.
17. W. Fischler, H. Nilles, J. Polchinski, S. Ray, K. Susskind, Phys. Rev. Lett. 47 (1981) 757.
18. P.S. Howe and P.C. West, in preparation.
19. E. Bergshoeff, M. de Roo and B. de Wit, Nucl. Phys. B182 (1981) 173;
    J.G. Taylor, in 'Superspace and Supergravity', eds. S.W. Hawking and M. Rocek, (C.U.P. 1981) 363.

20. P.S. Howe, Phys. Lett. 100B (1981) 389.
21. E. Sokatchev, Phys. Lett. 100B (1981) 466.
22. E. Fradkin and Tsetylin, Nucl. Phys. B203 (1982) 157.
23. P.S. Howe and U. Lindström, in 'Superspace and Supergravity' eds. S.W. Hawking and M. Rocek (C.U.P. 1981) 413; Nucl. Phys. B181 (1981) 487.
24. R.E. Kallosh, Phys. Lett. 99B (1981) 122.
25. P.S. Howe, K.S. Stelle and P.K. Townsend, Nucl. Phys. B191 (1981) 445.

UV SMOOTHNESS OF N = 4 EXTENDED SUPERSYMMETRIC YANG-MILLS FIELDS

R. Flume

Physikalisches Institut, Universität Bonn

ABSTRACT

We discuss the non renormalization of N=4 extended supersymmetric Yang-Mills fields in the framework of a Lorentz covariant component (Wess-Zumino) gauge.

I. The ultraviolet finiteness of N = 4 extended supersymmetric Yang-Mills fields has been established by Mandelstam [1] and by Brink, Lindgren and Nilsson [2] (see also the contribution of Brink to this conference and references quoted therein). These authors develop a new type of superfield formalism based on a light cone gauge, in which part of the N = 4 extended supersymmetry is realized linearly on the physical (i. e. propagating) field components. An alternative approach to display the non-renormalization of the N = 4 theory in the more conventional framework of a Lorentz covariant Wess-Zumino (WZ) gauge was proposed in [3]. I will outline in this contribution the strategy followed in ref. [3].

The main motivation to work in a Lorentz covariant WZ gauge - with its disadvantage of an entirely non-linear realization of supersymmetry - is that this gauge is presumably specially suited for a constructive attempt towards a conformal invariant solution of the model. Another more down to earth motivation is suggested for the following reason. All supersymmetric (susy) gauge theories handled with supergraph techniques - and in particular N = 4 extended Yang-Mills (Y.M.) fields in the susy light cone gauge - are difficult to control in the infrared, even on the level of off-shell Green's functions*). On the other hand there exist (at least off shell) no infrared problems in the WZ gauge. Susy Y M theories in the WZ gauge may be considered in this context as ordinary gauge theories with special renormalizable couplings to some matter fields. The off-shell infrared analysis of the latter models has been completed some years ago. (for a review cf. ref. [4]).

I will advocate the use of the renormalization formalism developed by Epstein and Glaser (EG) [5] as a substitute for a procedure involving a susy regularization. The crucial feature of the EG formalism, which will also be the essential ingredient in the discussion below, consists in its recursive nature. This will allow to push the non-linear susy Ward identities step by step into higher orders of perturbation theory. A so far unsolved problem arises however from those contributions to the susy Ward identities which are induced by the gauge fixing part of the Lagrangian, breaking supersymmetry. I have not found an efficient method to deal with these extra terms. May be, that one or another version of stochastic quantization (e.g. a decent Nicolai map) dispensing from gauge fixing will improve

*) The infrared problem of susy gauge theories in susy gauges has, to my best knowledge, not been solved so far.

the situation. For the time being I am forced to introduce an ad hoc hypothesis on the renormalization of N = 1 (non-extended) susy Y M fields. The hypothesis (introduced below) may be taken bona fide, if one believes in the equivalence of the N = 1 WZ component gauge and other supersymmetry preserving gauges. The desired u v properties have been proven for the latter case of susy gauges by all standards of rigour (cf. [6]). Taking the hypothesis on the renormalization of the N = 1 theory for granted, I will show that the step from N = 1 to N = 4 Y M fields can be accomplished rather easily.

The following section II contains an introduction to the EG formalism. In section III, I discuss the renormalization of N = 4 Y M fields.

II. In order to illustrate the formalism of Epstein and Glaser I discuss with their methods as a simplest possible example the renormalization of the one loop graph (fig. 1) appearing in a φ^4 theory. The formal ana-

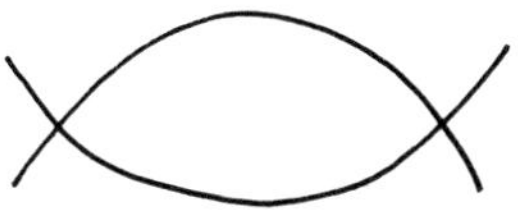

Fig. 1

lytic expression in x-space corresponding to the graph of Fig. 1 (with amputated external legs) is

$$\begin{aligned} &\langle T(:\varphi^2:(x) :\varphi^2: (y))\rangle \\ &= -2 \ \Delta^2_F (x-y), \\ &\Delta_F (x-y) = \frac{1}{\sqrt{2\pi}} \int \frac{d^d p}{p^2 - m^2 + i\epsilon} e^{ipx} \end{aligned} \quad (1)$$

where $\langle T(\ldots)\rangle$ denotes the time ordered free field vacuum expectation value of two normal ordered monomials $:\varphi^2:$. The time ordered function can be expressed through a function without ordering, an absorptive part, and a retarded commutator as follows:

$$\langle T(A(x)A(y)\rangle = \langle A(x)A(y)\rangle - \theta(y^0 - x^0)\langle [A(x), A(y)]\rangle, \tag{2}$$

where $A(x) = :\varphi^2:(x)$.

The absorptive part $\langle A(x)A(y)\rangle$ as well as the commutator function $\langle [A(x), A(y)]\rangle$ are well defined amplitudes as convolutions of free field Wightman functions. From the point of Eq. (2) the problem of renormalization is in this case to give a sense to the product of the two generalized function $\theta(y^0 - x^0)$ and $\langle [A(x), A(y)]\rangle$. The support of the commutator function is, due to locality, contained in the light cone $(x-y)^2 \geqslant 0$. So, in order to give a definite meaning to the product, that is to construct the retarded commutator function, one has to find an R(x - y) such that

$$R(x-y) = \begin{cases} \langle [A(x), A(y)]\rangle & \text{for } y^0 - x^0 \geqslant 0; \\ 0, & \text{otherwise.} \end{cases}$$

Pictorially speaking one has "to cut" the commutator at the point x = y, the tip of the light cone. To perform the cutting, one needs information about the singularity structure of the commutator function in the neighbourhood of the point x = y . It turns out that the superficial degree of divergence of the given diagram is (also in the general case) all one needs. For diagrams with a superficial degree of divergence < 0 the cutting procedure renders an unambiuous result+). The cutting of a diagram with a divergence degree $d \geqslant 0$ requires the specification of (d + 1) subtractions+). One has for the case at hand (Fig. 1) taken in four dimensions d = 0 and therefore one subtraction which is the one loop charge renormalization. I will not give the details of the cutting procedure but refer

---

+) This is true, if one does not generate extra singularities through oversubtractions.

to the original EG paper [5].

After this simple illustrating example I turn to the construction recipe for a general Feynman graph. Let $\langle T_F^{(-)}(X, Y)\rangle$ denote the time ordered (T) or anti-time ordered ($\bar{T}$) analytic expression corresponding to a Feynman graph F with external indices $X = (x_1 \ldots x_M)$ and internal vertices $Y = (y_1 \ldots y_N)$. Let the external propagators by amputated. We first define generalized absorptive parts of $\langle T(X, Y)\rangle$ denoted as $\langle A_F(x, Y)\rangle$ and $\langle D_F(x, Y)\rangle$ by

$$\langle A_F(x_1; X', Y)\rangle = \sum_{\substack{J \cup \bar{J} = \{X' \cup Y\} \\ J \cap \bar{J} = \phi \\ J \neq \phi}}^{(F)} (-1)^{|J|+1} \langle \bar{T}(J)\, T(\bar{J}, x_1)\rangle \tag{3}$$

where $X' = \{x_2, \ldots x_M\}$.

$$\langle D_F(x_1; X', Y)\rangle = \sum_{\substack{J \cup \bar{J} = \{X' \cup Y\} \\ J \cap \bar{J} = \phi \\ J \neq \phi}}^{(F)} \langle [T(\bar{J}, x_1), \bar{T}(J)]\rangle \cdot (-1)^{|J|} \,. \tag{4}$$

The sum $\sum^F$ in (3) and (4) goes over all subdivisions of $X' \cup Y$ into two subsets J and $\bar{J}$, with the quoted restrictions. Propagators connecting vertices in the subsets $(x_1, \bar{J})$ or J are to be taken as time ordered and anti-time ordered resp. (that is with a $+i\epsilon$ or $-i\epsilon$ prescription in momentum space). Propagators connecting the two sets of vertices $(x_1, \bar{J})$ and J are replaced by free field Wightman functions. The EG formalism is an inductive procedure in the number of vertices of the Feynman graphs. In this spirit we make the assumption that all Feynman graphs with less than (N + M) vertices have already been defined. $\langle A_F(\ldots)\rangle$ and $\langle D_F(\ldots)\rangle$ are then also well defined generalized functions. In fact, the components T and $\bar{T}$ appearing in (3) and (4) have less than (N + M) vertices and the convolution with the connecting Wightman functions, representing intermediate mass shell states, can be performed without problems. An important property of $\langle D(x; X', Y)\rangle$ (the generalization of the commutator function in Eq. (2)) is that its distribution theoretic support is contained in a multiple double cone:

$$\begin{aligned} &\text{support } \langle D_F(x_1; X', Y)\rangle \\ &= \{ (X, Y) \in R^{4(N+M)}; \\ &\quad (x_1 - x_j)^2 \geq 0 \quad (1 < j \leq M), \\ &\quad (x_1 - y_j)^2 \geq 0 \quad (1 \leq k \leq N)\}. \end{aligned} \tag{5}$$

(5) is a consequence of locality (cf. [5]). One has to find , as in the previous example, a retarded function characterised through the equation

$$\langle R_F(x_1; X', Y)\rangle = \begin{cases} -\langle D_F(x_1; X', Y\rangle \text{ for} \\ (x_j^{(0)} - x_1^{(0)}) \leq 0,\ (y_k^{(0)} - x_1^{(0)}) \leq 0, \\ 1 < j \leq M,\ 1 \leq k \leq N; \\ 0, \qquad \text{otherwise.} \end{cases}$$

$\langle T_F(X, Y)\rangle$ is then given in terms of $\langle R_F\rangle$ and $\langle A_F\rangle$:

$$\langle T_F(X, Y)\rangle = \langle R_F(x_1; X', Y)\rangle + \langle A_F(x_1; X', Y)\rangle.$$

To obtain the Green's functions one has finally to integrate the internal vertices:

$$\langle T_F(X) \rangle \equiv \int dY \langle T_F(X,Y) \rangle.$$

The convergence of the integrals in the case of a massless theory is a subtle question of the infrared structure which I will not go into.

Suppose one wants to establish some Ward identities in the renormalized theory. To achieve this one has to push the Ward identities through the induction machinery. So one takes as part of the inductive hypothesis that the Ward identities are satisfied by all Green's function of smaller than (N + M)-th order in perturbation theory. The same holds then automatically also for generalized absorptive parts $\langle D \rangle$ and $\langle A \rangle$ to order (N + M) (with symbols T replaced everywhere by D and A resp.) Indeed, the factors T and $\overline{T}$ appearing in (3) and (4) being of lower than (N + M)-th order satisfy separately the Ward identities by assumption. The connecting Wightman functions do not destroy anything since they represent the exchange of on-shell particles. The cutting of D-functions with a negative degree of divergence is also unproblematic. The u.v behaviour is in this case, roughly speaking, so smooth that it cannot endanger the Ward identities. The only point requiring careful examination is the cutting of D functions with a divergence degree larger or equal to zero. An example of the sort of consistency argument one needs is the content of the following section III. The efficiency of the EG formalism for an inductive proof of Ward identities should now be obvious. One has only to control a single step, the transition from the generalized absorptive part $\langle D(\ldots) \rangle$ to the retarded function $\langle R(\ldots) \rangle$. A discussion of overlapping divergencies is not needed.

I want to add a comment on the use of susy Ward identities. Those are first stated as relations among the generalized functions $\langle T(X,Y) \rangle$ including terms with total derivative acting on the internal vertices. Such terms are assumed to drop out on integration in the following. If they really vanish is again a question of the infrared structure, which is beyond the scope of this contribution.

III. The Lagrangian of N = 4 extended susy Y M fields may be considered as a special case of a N = 1 susy Lagrangian with attached chiral matter fields in the adjoint representation of the gauge group [7]. In the Wess-Zumino gauge it reads:

$$\mathcal{L} = \mathcal{L}_{I} + \mathcal{L}_{II} + \mathcal{L}_{III} , \tag{5}$$

$$\mathcal{L}_{I} = -\frac{1}{4} F_{\mu\nu}^{2} - i \bar{\chi} \bar{\sigma}^{\mu} D_{\mu} \chi ,$$

$$F_{\mu\nu} = \partial_{\mu} V_{\nu} - \partial_{\nu} V_{\mu} + g V_{\mu} \otimes V_{\nu} , \tag{5a}$$

$$D_{\mu} \chi = (\partial_{\mu} + g V_{\mu} \otimes) \chi ,$$

with the notation $(X \otimes Y)_a = f_{abc} X^b Y^c$ where f_{abc} are the structure constants of the gauge group. g is a coupling constant. Gauge group indices will always be suppressed. L_I represents the pure N = 1 vector multiplet Lagrangian in the Wess-Zumino gauge. χ denotes a Weyl field multiplet in the adjoint representation of the gauge group

$$\begin{aligned} \mathcal{L}_{II} = & \sum_{l=1}^{3} \{ i D_{\mu} \bar{\psi}_l \bar{\sigma}^{\mu} \psi_l - D_{\mu} \phi_l^{*} D^{\mu} \phi_l \} \\ & - \frac{g}{\sqrt{2}} (\bar{\psi}^i \phi^{*j} \otimes \bar{\psi}^k \epsilon_{ijk} + h.c.) \\ & - \frac{g^2}{2} \epsilon_{ijk} \phi^j \otimes \phi^k \epsilon_{ij'k'} \phi^{*j'} \otimes \phi^{*k'} . \end{aligned} \tag{5b}$$

$$\mathcal{L}_{III} = -g\sqrt{2} (\bar{\chi} \phi^i \otimes \bar{\psi}^i + h.c.) + \frac{g^2}{2} (\Sigma \phi^i \otimes \phi^{i*})^2 . \tag{5c}$$

L_{II} represents the chiral matter field Lagrangian of three multiplets (ϕ^i, ψ^i), each of them carrying the adjoint representation of the gauge group. L_{III} denotes the additional interaction, enforced by N = 1 supersymmetry, mixing the fields of the vector and chiral multiplets. The manifest invariances of the Lagrangian (5) are N = 1 supersymmetry, ordinary gauge invariance, SU(3) "flavour" symmetry of the matter fields and a chiral U(1) invariance. The notations have to be reorganized in order to display the larger SU(4) invariance (SU(3) x U(1) $\subset$ SU(4)), characterizing the N = 4 extended theory. The three chiral spinors ψ_i are grouped together with the vector multiplet spinor χ into a fundamental SU(4) multiplet:

$$(\psi^1, \ldots, \psi^3, \chi) \equiv (\lambda_1, \ldots, \lambda_4),$$

$$(\bar\psi^1, \ldots, \bar\psi^3, \bar\chi) \equiv (\bar\lambda^1, \ldots, \bar\lambda^4).$$

The scalars are arranged into the σ and $\bar\sigma$ representation of SU(4). Let η and $\bar\eta$ denote the O(4) tensors projecting into the (1,0) and (0,1) representation of O(4). Define

$$\phi_{li} = -\frac{1}{\sqrt{2}} \sum_a (\eta_{ali} A^a - i\, \bar\eta_{ali} B^a),$$

where

$$A^a = \sqrt{2}\, \mathrm{Re}\, \phi^a, \qquad B^a = \sqrt{2}\, \mathrm{Im}\, \phi^a,$$

and

$$\phi^{li} \equiv \phi^*_{li} = \frac{1}{2} \epsilon^{likm} \phi_{km}.$$

With the new notations the Lagrangian (5) reads as

$$\begin{aligned} \mathcal{L} = &-\frac{1}{4} F^2_{\mu\nu} - i\, \bar\lambda^i \bar\sigma^\mu D_\mu \lambda_i \\ &-\frac{1}{4} D_\mu \phi_{il} D^\mu \phi^{il} - \frac{g}{\sqrt{2}} (\bar\lambda^i \phi_{ik} \otimes \bar\lambda^k + h.c.) \\ &-\frac{g^2}{32} \phi_{ij} \otimes \phi_{kl}\, \phi^{ij} \otimes \phi^{kl}. \end{aligned} \tag{6}$$

This manifestly SU(4) invariant Lagrangian has been obtained by Brink, Scherk and Schwarz [8] through dimensional reduction. We quote for further reference the infinitesimal supersymmetry transformations which are Noether symmetries of the Lagrangian (6) (or (5)):

$$\begin{aligned} \delta_{\bar\xi, l} \bar\lambda^i &= -\frac{1}{\sqrt{2}} \delta_l^i \bar\xi \frac{\bar\sigma^{[\mu} \sigma^{\nu]}}{4} F_{\mu\nu} - \frac{1}{\sqrt{2}} \phi^{ik} \otimes \phi_{lk} \bar\xi, \\ \delta_{\bar\xi, l} \lambda_i &= -i\, \bar\xi \bar\sigma^\mu D_\mu \phi_{li}, \end{aligned} \tag{7}$$

$$\delta_{\bar{\xi},\ell}\, \phi_{ik} = \epsilon_{\ell ikj}\, \bar{\xi}\, \bar{\lambda}^{j},$$
$$\delta_{\bar{\xi},\ell}\, V^{\mu} = -\frac{e}{\sqrt{2}}\, \lambda_{\ell}\, \sigma^{\mu} \bar{\xi}.$$

$\bar{\xi}$ denotes here an anticommuting Weyl spinor parametrizing the variation. I will consider separately the susy transformations (7) and their hermitian conjugates. I will work in the Feynman gauge adding to (6) the gauge fixing piece

$$\mathcal{L}_{g.f.} = -\frac{1}{2}\, (\partial_{\mu} V^{\mu})^2 - \partial_{\mu} \bar{c}\, D^{\mu} c \tag{8}$$

where c and $\bar{c}$ denote Faddeev–Popov ghosts. The susy Ward identities are given by

$$0 = \langle T\, \delta^{(\ell)} (X) \rangle + \langle T\, s(X) \int \Delta_{(\ell)}\, d^4x \rangle, \tag{9}$$

where $\Delta_{(\ell)} = \frac{1}{\sqrt{2}}\, \bar{\xi}\, \bar{\sigma}^{\mu}\, \lambda_{\ell}\, \partial_{\mu} \bar{c}$.

X denotes an arbitrary bunch of fields $V_{\mu}, \chi, \bar{\chi}, \psi, \bar{\psi}, \phi$ and ϕ^*. The first term in (9) is the regular susy Ward identity corresponding to the susy transformation (7). I will use in the following mainly N = 1 terminology choosing in (9) $\ell = 4$ and making the identifications $\lambda_4 = \chi$; $\lambda_1, .., \lambda_3 = \psi^1, .., \psi^3$; $\phi_{4i} = \phi^i$.

The second term in (9) is induced through the gauge fixing Lagrangian (8). s(X) denotes the Slavnov variation of X:

$$s(V_{\mu}) = D_{\mu} c, \qquad s\psi = c \otimes \psi \qquad \text{etc.}$$

I have to introduce, as already mentioned in the introduction, an ad hoc hypothesis concerning the Lagrangian (5) considered as an N = 1 theory: The Lagrangian (5) can be renormalized with a minimal number of subtractions in conformity with gauge invariance, SU(3) flavour symmetry and the N = 1 susy Ward identities (Eq. 9 with $\ell = 4$) with the possible exclusion of the two-point function relations. The matter Yukawa couplings in (5b) remain unrenormalized.

Some remarks may be useful.

a) It has been shown by an explixit one loop calculation [9] in a fixed covariant WZ gauge that the wave function renormalizations (w.f.r.) are in fact not supersymmetric. But w.f.r.'s for themselves are irrelevant.

What counts are the relations of the w.f.r.'s to the vertex corrections. The pure vector field Lagrangian L_I (5a), in particular, is completely characterized through gauge invariance and masslessness of the Weyl spinors χ. These two principles imply N = 1 supersymmetry for L_I. Part of the general hypothesis is that w.f.r.'s are fixed relative to the vertex corrections so that N = 1 supersymmetry is perserved.

b) The hypothesis on minimal subtractions comprises the assumptions that all composite operators appearing in the susy Ward identities (9) are renormalized according to their canonical dimensions. The only other information needed about the composite operators is that they do not contribute to on shell amplitudes. This is used for establishing the Ward identities of the generalized absorptive parts D and A: The susy Ward identities are supposed to linearize on the intermediate massless states connecting the different time and anti-time ordered components of D and A resp..

c) Working with a susy guage fixing and supergraphs it is a matter of simple power counting to verify that the radiative corrections to the Yukawa coupling of the chiral matter fields remain unrenormalized [10]. An alternative formal argument for the same purpose avoiding supergraph techniques, can be made also in a covarian WZ gauge. The proof for this argument is unfortunately plagued by the same (unsolved) problems as the general renormalization proof of N = 1 supersymmetry since the unpleasant correction terms induced through the gauge fixing Lagrangian interfere also here.

I want to demonstrate that on the basis of the preceding N = 1 hypothesis it is rather straightforward to settle the extended N = 4 supersymmetry. I go for this purpose through the list of those potentially u.v. divergent Grenns' functions which determine the fate of the crucial O(4) invariance *) of the N = 4 extended theory.

The postulated SU(3) flavour symmetry implies for the functions $\langle T\phi_i^*\phi_i\rangle$, $\langle T\phi_i^*\phi_i V_\mu\rangle$ and $\langle T\phi_i^*\phi_i V_\mu V_\nu\rangle$ the desired O(4) invariance. In fact, in this context it is only a question of changing the notation from $\phi_i\,(\phi_i^*)$ to $\phi_{\ell i}\,(\phi^{\ell i})$ to make the O(4) invariance manifest. Next I consider scalar four point functions $\langle T\phi_i\phi_j\phi_k^*\phi_l^*\rangle$.

*) It is sufficient for our purposes to consider the real SO(4) subgroup of SU(4).

They appear in the N = 1 susy Ward identity

$$
\begin{aligned}
0 = & - \langle T \phi_i \, i \bar{\xi} \bar{\sigma}^\mu D_\mu \phi_j \phi_k^* \phi_l^* \rangle \\
& + \langle T \phi_i \psi_j \bar{\xi} \bar{\psi}_k \phi_l^* \rangle + \langle T \phi_i \psi_j \phi_k^* \bar{\xi} \bar{\psi}_l \rangle \\
& + \frac{1}{\sqrt{2}} \langle T s (\phi_i \psi_j \phi_k^* \phi_l^*) \int \Delta_{(4)} d^4x \rangle .
\end{aligned} \tag{10}
$$

All terms in (10) besides the linear piece $- \langle T \phi_i \, i \bar{\xi} \bar{\sigma}^\mu \partial_\mu \phi_j \phi_k^* \phi_l^* \rangle$ of the first term are superficially u.v. convergent. It follows that one has not to specify an independent subtraction prescription for the scalar four point functions. The u.v. structure of $\langle T \phi^* \phi^* \phi \phi \rangle$ is governed by the irreducible, potentially u.v. divergent subdiagrams of the other, superficially convergent terms in (10). The net conclusion therefore is that O(4) invariance on the level of four point functions will be guaranteed if it is settled for two and three point functions.

I turn now to the central discussion of the three and two point functions. Suppose that O(4) invariance has been established up to n-th order in the coupling constant for all Greens' functions. One consequence is that the chiral three point functions $\langle T \bar{\psi} \bar{\psi} \phi^* \rangle$ and the three point functions $\langle T \bar{\chi} \bar{\psi} \phi \rangle$ induced by the vector multiplet can be assembled to this order into a O(4) covariant set of functions*). The Greens' functions $\langle T \bar{\psi} \bar{\psi} \phi^* \rangle$ are by hypothesis non-renormalized to all orders in perturbation theory. The same holds then up to n-th order for $\langle T \bar{\chi} \bar{\psi} \phi \rangle$. O(4) invariance carries over to the generalized absorptive parts $\langle D(\bar{\psi} \bar{\psi} \phi^*) \rangle$, $\langle D(\bar{\chi} \bar{\psi} \phi) \rangle$ and $\langle A(\bar{\psi} \bar{\psi} \phi^*) \rangle$, $\langle A(\bar{\chi} \bar{\psi} \phi) \rangle$ in (n + 1)-th order. The u.v. behaviour of $\langle D(\bar{\psi} \bar{\psi} \phi) \rangle$ is because of the non-renormalization hypothesis so smooth that the transition to the retarded function $\langle R(\bar{\psi} \bar{\psi} \phi) \rangle$ and therewith to $\langle T(\bar{\psi} \bar{\psi} \phi) \rangle$ is unique that is subtraction free. O(4) covariance can then be used to find also a subtraction free definition of $\langle R(\bar{\chi} \bar{\psi} \phi) \rangle$ and $\langle T(\bar{\chi} \bar{\psi} \phi) \rangle$ resp. in (n + 1)-th order. It has to be checked whether this O(4) covariant and subtraction free definition of

---

*) The same holds for the hermitean conjugate set of functions $\langle T \psi \psi \phi \rangle$ and $\langle T \chi \psi \phi^* \rangle$.

$\langle T(\bar\chi\bar\psi\phi)\rangle$ is compatible with N = 1 supersymmetry. A special N = 1 susy Ward identity involving $\langle T(\bar\chi\bar\psi\phi)\rangle$ is given by

$$-\frac{1}{\sqrt{2}}\langle T\bar\xi\left(\frac{\bar\sigma^{[\mu}\sigma^{\nu]}}{4}F_{\mu\nu}+\sum_{\ell}\phi^{*\ell}\otimes\phi^{\ell}\right)\bar\psi^{i}\psi^{i}$$

$$-\frac{1}{\sqrt{2}}\langle T\bar\chi\,\epsilon_{ijk}(\phi^{j}\otimes\phi^{k})\psi^{i}\rangle-i\langle T\bar\chi\bar\psi^{i}\xi\bar\sigma^{\mu}D_{\mu}\phi^{i}\rangle \qquad (11)$$

$$+\frac{1}{\sqrt{2}}\langle T s(\bar\chi\bar\psi_i\psi_i)\int\Delta_{(4)}\,d^4x\rangle=0.$$

Assuming that (11) is satisfied up to nth order in perturbation theory I perform once more the inductive step to the (n+1)-th order. All terms in (11) besides the linear pieces $\frac{1}{\sqrt{2}}\langle T\bar\xi\frac{\sigma^{[\mu}\sigma^{\nu]}}{4}\partial_{[\mu}V_{\nu]}\bar\psi^{i}\psi^{i}\rangle$ and $-i\langle T\bar\chi\bar\psi^{i}\xi\bar\sigma^{\mu}\partial_{\mu}\phi\rangle$ of the first and third bracket are superficially u.v. convergent and therefore, invoking the minimal subtraction hypothesis, uniquely determined also in (n + 1)-th order. Since $\langle T\bar\chi\bar\psi\phi\rangle$ has already been fixed the only possibility to satisfy (11) is to read it as a definition for $\langle T\xi\bar\sigma^{[\mu}\sigma^{\nu]}\partial_{[\mu}V_{\nu]}\bar\psi^{i}\psi^{i}\rangle$. (This of course determines also $\langle TV_\mu\bar\psi\psi\rangle$ to (n + 1)-th order). Eq.(11) is satisfied so by construction. The vertex function $\langle TV_\mu\bar\chi\chi\rangle$ is given together ith $\langle TV_\mu\bar\psi\psi\rangle$ through O(4) invariance. It cannot be excluded that the now - with the criterion of O(4) invariance and Eq. (11) - constructed (N + 1)-th order Greens' functions $\langle TV_\mu\bar\psi\psi\rangle$, $\langle TV_\mu\bar\chi\chi\rangle$ and $\langle T\bar\psi\bar\chi\phi\rangle$ deviate from the renormalized Greens' functions, as they are postulated for the Lagrangian (5) regarded as a N = 1 susy theory. To distinguish the two sets of Greens' functions I add to the former a subscript C and to the latter a subscript P. I want to show that the possible deviations $\langle T\rangle_C-\langle T\rangle_P$ are compatible with N = 1 supersymmetry. One has first to note that gauge invariance requires that with the vertex functions $\langle TV_\mu\bar\psi\psi\rangle_C$ and $\langle TV_\mu\bar\chi\chi\rangle_C$ also the w.f.r.'s governing $\langle T\bar\psi\psi\rangle_C$ and $\langle T\bar\chi\chi\rangle_C$, should be O(4) invariant. We take as defining relations for the spinor two point functions

$$\langle T\bar\psi\psi\rangle_C=\langle T\bar\psi\psi\rangle_P$$

and

$$\langle T\bar\chi\chi\rangle_C=\langle T\bar\psi\psi\rangle_C.$$

It can be that $\langle T\bar{\chi}\chi\rangle_P$ is unequal to $\langle T\bar{\chi}\chi\rangle_C$. The possible modification of $\langle T\bar{\chi}\chi\rangle_C$ compared to $\langle T\bar{\psi}\psi\rangle_P$ is achieved through a finite change of the χ--w.f.r. Both sets $\langle TV_\mu\bar{\psi}\psi\rangle_C$, $\langle T\bar{\chi}\bar{\psi}\phi\rangle_C$, ... and $\langle TV_\mu\bar{\psi}\psi\rangle_P$, $\langle T\bar{\chi}\bar{\psi}\phi\rangle_P$ ... satisfy the Ward identity (11), the former by construction and the latter by hypothesis. Since the deviations (if there are any) of the functions $\langle T\rangle_C$ from the functions $\langle T\rangle_P$ are assumed to occur only at the step from n-th to (n + 1)-th order of perturbation theory, the differences $\langle T\bar{\psi}\psi V_\mu\rangle_C - \langle T\bar{\psi}\psi V_\mu\rangle_P$ and $\langle T\bar{\chi}\bar{\psi}\phi\rangle_C - \langle T\bar{\chi}\bar{\psi}\phi\rangle_P$ are generated by a finite counterterm, which, because of (11) has the form

$$\Delta\mathcal{L} = \delta g\left(-i\bar{\psi}_i\bar{\sigma}^\mu V_\mu \otimes V_i - \frac{1}{\sqrt{2}}\bar{\chi}\,\phi_i \otimes \bar{\psi}_i\right). \tag{12}$$

ΔL is finally absorbed into a redefinition of the coupling constant:

$$\mathcal{L}_P(g) \rightarrow \mathcal{L}_C = \mathcal{L}_P(g+\delta g).$$

Following this procedure we preserve gauge invariance (with a possibly modified gauge coupling) and therefore also N = 1 supersymmetry for L_I, (5a). O(4) invariance is imposed by construction. N = 1 supersymmetry for the chiral matter field Lagrangian $L_{II} + L_{III}$ ((5b) + (5c)), is also guaranteed as the difference Δ L, (12) has been promoted to an overall change of the coupling constant. The induced Yukawa couplings $\bar{\chi}\phi\otimes\bar{\psi}$ inherit from the chiral Yukawa coupling $\bar{\psi}\phi^*\otimes\bar{\psi}$ by O(4) invariance the property to remain unrenormalized. The O(4) invariance (and there with also u.v. finiteness) of the ϕ^4 couplings is implied by the O(4) invariance of the Yukawa couplings and the w.f.r.'s. One consequence is, as already noted by Stelle and Townsend [11], that the renormalization group β function vanishes. I should mention, however, that the last conclusion cannot be deduced directly from the above discussion. One has to appeal for this purpose to the equivalence of the EG formalism with other more conventional renormalization procedures. It is hard to keep track of the renormalization group within the EG formalism.

References

[1] S. Mandelstam, Nucl. Phys. B213 (1983) 149.

[2] L. Brink, O. Lindgren and B.E.W. Nilsson, Nucl. Phys. B212 (1983) 401.

[3] R. Flume, Nucl. Phys. B217 (1983) 531.

[4] P.K. Mitter, in New Developments in Quantum Theory and Statistical Mechanics, M. Lévy and P.K. Mitter edit.s, Plenum Press, New York (1977).

[5] H. Epstein and V. Glaser, Ann. Inst. Henri Poincaré, vol. XIX no. 3 (1973) 211.

[6] C. Piguet and K. Sibold, B197 (1982) 257, B197 (1982) 257.

[7] P. Fayet, Nucl. Phys. B149 (1979) 137.

[8] L. Brink, J.H. Schwarz and J. Scherk, Nucl. Phys. B121 (1977) 77.

[9] R. Barbieri, S. Ferrara, L. Maiani, F. Palumbo and C.A. Savoy, CERN-preprint, Ref. TH 3282-CERN.

[10] S. Ferrara and P. Piguet, Nucl. Phys. B93 (1975) 261; M.T. Grisaru, W. Siegel and M. Rocek, Nucl. Phys. B159 (1979) 429.

[11] K. Stelle, Proceedings of High Energy Conference 1982, Imperial College Preprint.

THE GEOMETRY OF N=2 SUPERSYMMETRY

German Sierra

Laboratoire de Physique Théorique de l'Ecole Normale Supérieure
24 rue Lhomond, 75231 Paris cedex 05
FRANCE

Scalar fields play a very important role in Supersymmetry. They belong to matter supermultiplets for N=1 and N=2 and to vector multiplets for N=2. For $N > 4$ they are always present in supergravity multiplets. The coupling to supergravity induces "non-linearities" in the scalar sector which we must guess in advance in order to construct the theory to all orders in K (Newton's constant). One must conjecture the coset space G/H in which the scalars sit, and then prove that it gives a consistent theory. In rigid supersymmetry we are not forced to consider these kinds of non linearities in the sense that the dimensional scalar self-coupling constant is unrelated to the Newton constant. However, if we do consider non linearities then we can see which are the restrictions implied by supersymmetry and subsequently investigate the effect of supergravity.

To begin with let us consider a set of free massless scalar fields ϕ^x (x=1,...n) whose Lagrangian is simply given by

$$\mathcal{L}_0 = \frac{1}{2} \partial_\mu \phi^x \partial^\mu \phi^x \quad . \tag{1}$$

This Lagrangian is invariant under SO(n) rotations of ϕ^x.

If we want to deal with more general transformations of the scalar fields

$$\delta\phi^x = \xi^x(\phi) \tag{2}$$

the $\xi^x(\phi)$ being arbitrary functions of ϕ^x, then we must consider Lagrangians of the form

$$\mathcal{L}_\sigma = \frac{1}{2} g_{xy}(\phi) \partial_\mu \phi^x \partial^\mu \phi^y \tag{3}$$

(σ stands for non linear σ-models).

The variation of $\mathcal{L}_\sigma$ under an arbitrary transformation $\delta\phi^x = \xi^x(\phi)$ is equivalent to a variation of g_{xy} given by[1]

$$\delta g_{xy} = \xi^z g_{xy,z} + \xi^z{}_{,x} g_{zy} + \xi^z{}_{,y} g_{zx} \ . \tag{4}$$

As an example we can take the free case where $g^{(0)}_{xy} = \delta_{xy}$ and $\xi^x = c^x + \Lambda^x{}_y \phi^y$. Then eq.(4) is

$$\delta g_{xy} = \Lambda_{xy} + \Lambda_{yx} = 2\Lambda_{(xy)} \ . \tag{5}$$

The invariance of $\mathcal{L}_0$ requires $\delta g_{xy} = 0$ so that $\Lambda_{xy} = \Lambda_{[xy]}$ are the infinitesimal generators of SO(n), and c^x the translations. The invariance group of $\mathcal{L}_0$ is thus the semidirect product of T_n and SO(n).

This example suggests that we can regard g_{xy} as the metric of a Riemannian manifold $\mathcal{M}$ parametrized by the scalars ϕ^x . Eq.(4) is indeed a general coordinate transformation of the metric g_{xy} , and can also be written as

$$\delta g_{xy} = 2\,\xi_{(x;y)} \tag{6}$$

where the semicolon denotes covariant differentiation with respect to the Christoffel connection

$$\Gamma_{xy}{}^z = \frac{1}{2} g^{zw} \left(g_{wx,y} + g_{wy,x} - g_{xy,w} \right) . \tag{7}$$

This shows that the symmetries of the scalar Lagrangian $\mathcal{L}_\sigma$ are in one-to-one correspondence with the isometries of the manifold $\mathcal{M}$, which are generated by the Killing vectors ξ^x satisfying the Killing equation

$$\xi_{(x;y)} = 0 \tag{8}$$

It is interesting to note that the commutator of two Killing vectors is again a Killing vector as a consequence of the cyclic identity of the Riemann tensor ($R^x{}_{[yzw]} = 0$) . A complete set of Killing vectors form a group G in that

$$[\xi_{(r)}, \xi_{(s)}]^x \equiv \left[\xi^x_{(r),y}\, \xi^y_{(s)} - (r \leftrightarrow s) \right] = f_{rs}{}^t\, \xi^x_{(t)} \tag{9}$$

where $f_{rs}{}^t$ are the structure constants of G.

The physics described by these models depends exclusively on the scalar manifold $\mathcal{M}$ and not on the particular form adopted by the metric g_{xy} . If $\mathcal{M}$ is flat, no matter how complicated is g_{xy} , there will be no interactions and the S-matrix will be trivial.

In ordinary scalar theories the interactions are given by mass terms and potential terms. They may also have a geometrical interpretation, although less immediate. For instance the mass term $m^2 \phi_x^2$ can be generalized in the case of a manifold with a semisimple isometry group G as follows

$$\mathcal{L}_m = -\frac{m^2}{2}\, \xi^x_{(r)}\, \xi^y_{(s)}\, g_{xy}\, g^{rs} \tag{10}$$

where g^{rs} is the inverse of the Cartan-Killing metric. $\mathcal{L}_m$ is invariant under the group G, because f_{rst} is totally antisymmetric, for G semi-simple.

For arbitrary potential terms we do not expect to maintain the full G-invariance.

Of phenomenological interest are higher derivative Lagrangians of the form[2]

$$\mathcal{L}'_d = c_{x_1 x_2 \cdots x_d}(\phi)\, \partial_{\mu_1}\phi^{x_1}\, \partial_{\mu_2}\phi^{x_2} \cdots \partial_{\mu_d}\phi^{x_d}\, \varepsilon^{\mu_1 \cdots \mu_d} \tag{11}$$

in d-dimensional space-time, where $c_{x_1 \cdots x_d}$ is a total antisymmetric tensor defined on $\mathcal{M}$. When the manifold $\mathcal{M}$ has a semisimple group G of isometries, there exists a natural tensor

$$c_{xyz} = \xi^{(r)}_x\, \xi^{(s)}_y\, \xi^{(t)}_z\, f_{rst} \tag{12}$$

such that $\mathcal{L}'_3$ turns out to be invariant under the full group G as a consequence of the Jacobi identities satisfied by the f's.

It is also interesting to note that in two dimensions $\mathcal{L}'_2$ has two derivatives, as does $\mathcal{L}_\sigma$ so that the full kinetic Lagrangian in this case has the form

$$\mathcal{L} = \frac{1}{2}\, g_{xy}\, \partial_\mu\phi^x\, \partial^\mu\phi^y + c_{xy}\, \partial_\mu\phi^x\, \partial_\nu\phi^y\, \varepsilon^{\mu\nu}\,. \tag{13}$$

Another way of obtaining interaction terms is through non trivial dimensional reduction. For that purpose we must specify the dependence of the scalar fields ϕ^x on the extra coordinate(s), say x^d, such that the reduced action in d-1 dimensions is independent of x^d. Setting

$$\frac{\partial}{\partial x^d}\phi^x = \xi^x_{(r)}\, m^r \tag{14}$$

we obtain a consistent reduction if $\xi^x_{(r)}$ are Killing vectors of $\mathcal{M}$. This result generalizes the method of Scherk and Schwarz[3], who used only linear symmetries of the action.

The reduced Lagrangian in d-1 is therefore

$$\mathcal{L}_{d-1} = \frac{1}{2}\, g_{xy}\, \partial_\mu\phi^x\, \partial^\mu\phi^y - \frac{1}{2}\, m^r m^s\, \xi^x_{(r)}\, \xi^y_{(s)}\, g_{xy}\,. \tag{15}$$

We can ask under which conditions the isometry group G of $\mathcal{L}_d$ is still an invariance group of $\mathcal{L}_{d-1}$. Performing a transformation

$$\delta\phi^x = \xi^x_{(t)}\, c^t \tag{16}$$

we get

$$\delta\mathcal{L}_{d-1} = -m^r\, \xi_{(r)x}\, [m^s\, \xi_{(s)}, \xi_{(t)}\, c^t]^x\,. \tag{17}$$

Thus $\xi_{(t)} c^t$ must commute with $m^s \xi_{(s)}$ for $\delta\phi^x$ to be an invariance. The Killing vectors in (14) are usually taken in the Cartan subalgebra of G so that the number of independent mass parameters m^r equals the rank of G.

Applying this method to $\mathcal{L}'_3$ with C_{xyz} given by eq.(12), we obtain a Lagrangian of the same type in two dimensions

$$\mathcal{L}'_2 = 3 f_{rst} m_u \xi^{(r)}_x \xi^{(s)}_y \xi^{(t)}_z \xi^{(u)z} \partial_\mu \phi^x \partial_\nu \phi^y \epsilon^{\mu\nu} \tag{18}$$

which has the same invariance properties as $\mathcal{L}_{d-1}$.

By now the geometrical interpretation of the models we have been discussing is quite clear. Before continuing let us point out that the quantity $\partial_\mu \phi^x$ transforms as a vector under a general "coordinate" transformation $\delta\phi^x = \xi^x(\phi)$ viz.

$$\delta \partial_\mu \phi^x = \xi^x{}_{,y} \partial_\mu \phi^y = \xi^x{}_{,y} \partial_\mu \phi^y - \xi^y (\partial_\mu \phi^x)_{,y} \tag{19}$$

recalling that a vector v^x transforms as

$$\delta v^x = \xi^x{}_{,y} v^y - \xi^y v^x{}_{,y} . \tag{20}$$

This is equally true for $\partial \rightarrow D$, the spinor derivative of supersymmetry in the case that ϕ^x is a superfield.

The global symmetries of $\mathcal{M}$ can be made local by minimal coupling to gauge fields. Suppose we gauge a subgroup H of G, then the covariant derivative of ϕ^x will be given by[4]

$$(\mathcal{D}_\mu \phi)^x = \partial_\mu \phi^x + \xi^x_{(\hat{r})} A^{(\hat{r})}_\mu \tag{21}$$

where $\xi^x_{(\hat{r})}$ are the Killing vectors of H.

If we transform ϕ^x as

$$\delta \phi^x = \xi^x_{(r)} c^{(r)}(x) \tag{22}$$

and $\xi^x_{(\hat{r})}$ as indicated by (20) we get

$$\begin{aligned} \delta(\mathcal{D}_\mu \phi)^x = {} & c^{(r)} \xi^x_{(r),y} (\mathcal{D}_\mu \phi)^y - c^{(r)} \xi^y_{(r)} (\mathcal{D}_\mu \phi)^x{}_{,y} \\ & + (\delta A^{(\hat{t})}_\mu + \partial_\mu c^{(\hat{t})} + A^{(\hat{r})}_\mu c^{(s)} f_{\hat{r}s}{}^{\hat{t}}) \xi^x_{(\hat{t})} \\ & + (A^{(\hat{r})}_\mu c^{(\check{s})} f_{\hat{r}\check{s}}{}^{\check{t}} + \partial_\mu c^{(\check{t})}) \xi^x_{(\check{t})} \end{aligned} \tag{23}$$

where $\{\xi^x_{(\check{t})}\}$ is the set of Killing vectors of the coset G/H = K.

The first term in (23) is the covariant transformation law of $(\mathcal{D}_\mu \phi)^x$. The invariance under local transformations of H implies the transformation law for the gauge fields

$$\delta A^{(\hat{t})}_\mu = - (\mathcal{D}_\mu c)^{(\hat{t})} . \tag{24}$$

The G-invariance will be broken to H unless the third term in (23) vanishes, i.e.

$$f_{\hat{r}\check{s}}{}^{\check{t}} = 0 \quad , \quad \partial_\mu c^{(\check{t})} = 0 \quad . \tag{25}$$

Note that $f_{\hat{r}\hat{s}}{}^{\check{t}} = 0$ by the group property of $H([H,H] \subset H)$.

The equation $f_{\hat{r}\check{s}}{}^{\check{t}} = 0$ ($[H,K] \subset H$) will be satisfied if H is a <u>normal</u> subgroup of G [5]. In this case the gauge field $A_\mu^{(\hat{t})}$ also transforms under K as

$$\delta_K A_\mu^{(\hat{t})} = \left(f_{\check{s}\hat{r}}{}^{\hat{t}}\, c^{\check{s}}\right) A_\mu^{(\hat{r})} = K^{(\hat{t})}{}_{(\hat{r})}\, A_\mu^{(\hat{r})} \quad . \tag{26}$$

The Lagrangian of the gauge invariant σ-model is

$$\mathcal{L} = \frac{1}{2} g_{XY} (\mathcal{D}_\mu \phi)^X (\mathcal{D}_\mu \phi)^Y - \frac{1}{4} F_{\mu\nu}^{(\hat{r})} F_{\mu\nu}^{(\hat{r})} \quad . \tag{27}$$

Although eq.(27) seems quite general we can consider more complicated Lagrangians such as

$$\mathcal{L} = \frac{1}{2} g_{XY}(\phi)\, \partial_\mu \phi^X \partial_\mu \phi^Y - \frac{1}{4} a_{IJ}(\phi)\, F_{\mu\nu}^I F_{\mu\nu}^J \tag{28}$$

which describe the interaction of n scalar fields ϕ^X and m Maxwell fields ($F_{\mu\nu}^I = 2\, \partial_{[\mu} A_{\nu]}^I$).

A new feature is the appearance of a new metric $a_{IJ}(\phi)$ in the Maxwell action. This forces the Maxwell fields to transform under the isometry group G if the G-invariance is to be maintained.

Writing the transformation of $F_{\mu\nu}^I$ as

$$\delta F_{\mu\nu}^I = c^r\, \Lambda_{(r)}{}^I{}_J\, F_{\mu\nu}^J \tag{29}$$

we obtain the following equation for a simultaneous variation $\delta\phi^X = \xi_{(r)}^X c^r$ of the scalar fields

$$a_{IJ,X}\, \xi_{(r)}^X + 2\, a_{K(I}\, \Lambda_{(r)}{}^K{}_{J)} = 0 \tag{30}$$

Varying eq.(30) we get

$$[\Lambda_{(r)}, \Lambda_{(s)}]^I{}_J = f_{rs}{}^t\, \Lambda_{(t)}{}^I{}_J \tag{31}$$

which means that the vectors transform as a linear representation of the group G.

This situation arises in extended supergravity theories, i.e. for N=8 in d = 5 where G is the non compact group E_6 and the vectors belong to its 27 dimensional representation[6]. In d=4 the situation is slightly different because the non compact group E_7 is the invariance group of the equation of motions and not a symmetry of the action.

All this is valid for any theory with scalar and vector fields. As we have seen there is no restriction on the kind of scalar manifold we can have or on the gauge group. Our aim is to show that supersymmetry imposes severe restrictions on $\mathcal{M}$. Restrictions on the gauge group arise only in the context of supergravity and will not be treated here.

Briefly put : supersymmetry determines the holonomy group Γ of the scalar manifold according to the number of supersymmetries N in a given dimension. We will concentrate on matter supermultiplets where the number of scalars is the same for all dimensions. The fact that the holonomy group Γ of an n-dimensional real manifold $\mathcal{M}$ differs from SO(n) (the largest one), has enormous consequences for the geometry and topology of $\mathcal{M}$.

The infinitesimal generators of the holonomy group are given by the Riemann tensor $R_{xy}{}^a{}_b$ where $x, y, \ldots$ are "world indices" and $a, b, \ldots$ flat indices belonging to the tangent space. It is quite convenient to introduce a "vielbein" f^a_x, together with its inverse f^x_a, which allows us to define the spin connection $\omega_x{}^a{}_b$ as in general relativity

$$f^a_{[x,y]} + \omega_{[y}{}^a{}_b\, f^b_{x]} = 0 \quad . \tag{32}$$

The Riemann tensor is then written as

$$R_{xy}{}^a{}_b = \omega_y{}^a{}_{b,x} + \omega_x{}^a{}_c\, \omega_y{}^c{}_b - (x \leftrightarrow y) \, . \tag{33}$$

A flat space has trivial holonomy group. Since Γ is contained in the tangent space group T any restriction on T will hold for Γ.

The supersymmetry variation of ϕ^x gives us the fermions that are considered as vectors in the tangent space

$$\delta \phi^x = f^x_a \, (\epsilon\lambda)^a \quad . \tag{34}$$

Thus the tangent space group should contain the automorphism group of the supersymmetry algebra and the symmetry group acting on the fermions.

Let us take the case of N=1 in d=4. The transformation laws of the chiral field ϕ^a ($a = 1, \ldots, n/2$) and its complex conjugate $\bar{\phi}^a$ in the flat case

$$\delta\phi^a = \epsilon^\alpha \lambda^a_\alpha \qquad\qquad \delta\bar{\phi}^{\bar{a}} = \bar{\epsilon}_{\dot{\alpha}}\, \bar{\lambda}^{\dot{\alpha}\bar{a}} \tag{35}$$

can be written in the "curved" case following (34) as

$$\begin{aligned} \delta\phi^x &= f^x_a\, \epsilon^\alpha \lambda^a_\alpha + f^x_{\bar{a}}\, \bar{\epsilon}_{\dot{\alpha}} \bar{\lambda}^{\dot{\alpha}\bar{a}} \\ \delta\phi^{\bar{x}} &= f^{\bar{x}}_{\bar{a}}\, \bar{\epsilon}_{\dot{\alpha}} \bar{\lambda}^{\dot{\alpha}\bar{a}} + f^{\bar{x}}_a\, \epsilon^\alpha \lambda^a_\alpha \end{aligned} \tag{36}$$

where $f^x_a, f^x_{\bar{a}}, \ldots$ are the various components of the vielbein (with ϕ^x and $\phi^{\bar{x}}$ considered as independent variables).

Similarly the chiral constraint

$$\bar{D}_{\dot{\alpha}}\, \phi^a = 0 \tag{37}$$

now reads

$$f^a_x \, \bar{D}_{\dot\alpha} \phi^x \quad + \quad f^a_{\bar{x}} \, \bar{D}_{\dot\alpha} \phi^{\bar{x}} \quad = 0 \ . \tag{38}$$

It is easy to prove that eq.(38) is preserved assuming the tangent space group to be U(n/2) : the only non-vanishing components of the spin connection are $\omega_x{}^a{}_b, \omega_{\bar{x}}{}^a{}_b$ and its complex conjugate. Note that the U(1) factor corresponds to the automorphism group of supersymmetry while SU(n/2) is the symmetry group of the fermions. This kind of space is called Kahlerian and its connection with N=1 supersymmetry was first shown by Zumino[7]. In coupling the chiral multiplet to supergravity one gets Kahler manifolds of a restricted type (Hodge manifolds) where the scalar self-coupling constant is quantized in terms of the Newton constant[8].

For N=2 the scalar manifold $\mathcal{M}$ is hyper-Kahler as was proved by Alvarez-Gaumé and Freedman[9]. A hyper-Kahler manifold is a manifold whose holonomy group Γ is contained in Sp(n/4). To show that this is the case we will use superfield formalism in d=6 [10] in a formalism for which the automorphism group SU(2) of N=2 supersymmetry in d=4 is manifest (the U(1) is recovered through dimensional reduction d=6 → d=4).

The matter multiplet in d=6 is described by the hypermultiplet satisfying the reality condition

$$(\phi^{ia})^* = \epsilon_{ij} \, \Omega_{ab} \, \phi^{jb} \tag{39}$$

where Ω_{ab} is the simplectic metric of Sp(n/4).

The spinors in d=6 areof simplectic type and satisfy

$$(\lambda^a_\alpha)^* = B_{\dot\alpha}{}^\beta \, \lambda^b_\beta \, \Omega_{ba} \tag{40}$$

where $\alpha \in SU^*(4) \simeq SO(5,1)$.

Thus we do not need to consider the complex conjugate of λ^a_α . The supersymmetry algebra is given by

$$\{D^i_\alpha \, , \, D^j_\beta\} = i \, \epsilon^{ij} \, \partial_{\alpha\beta} \ . \tag{41}$$

The hypermultiplet is defined by the superfield constraint [5,11]

$$D^{(i}_\alpha \, \phi^{j)a} = 0 \tag{42}$$

whose solution is

$$D^i_\alpha \, \phi^{ja} = \epsilon^{ij} \, \lambda^a_\alpha \tag{43}$$

This is an on-shell multiplet : further spinor differentiation produces the equations of motion,

$$\partial^{\alpha\beta} \, \partial_{\alpha\beta} \, \phi^{ia} = 0 \quad , \quad \partial^{\alpha\beta} \, \lambda^a_\beta = 0 \ . \tag{44}$$

From eqs. (39) and (40) we see that the pair of indices (i,a) belong to the tangent space Sp(1) x Sp(n/4). The free case constraint (42) can be then generalized by introducing a vielbein f^{ia}_x. With the assumed factorization of the tangent space indices (i,a) eq.(42) reads :

$$f^{(ia}_x \, D^{j)}_\alpha \phi^x = 0 \tag{45}$$

which is solved by

$$D_{\alpha i} \phi^x = f^x_{ia} \lambda^a_\alpha \, . \tag{46}$$

The spin connection $\omega_x{}^{ia}{}_{jb}$ splits into two pieces :

$$\omega_x{}^{ia}{}_{jb} = \omega_x{}^i{}_j \delta^a_b + \omega_x{}^a{}_b \delta^i_j \tag{47}$$

where $\omega_x{}^i{}_j$ is the Sp(1) connection and $\omega_x{}^a{}_b$ the Sp(n/4) one. Supersymmetry implies further restrictions on $\mathcal{M}$. First of all, from (45) we deduce the "representation preserving constraint"

$$\omega_{a(ijk)} = 0 \, . \tag{48}$$

As in the free case, eq.(45) is an on-shell constraint and implies the equations of motion for ϕ^x and λ^a_α. These are derivable from an action if

$$\omega_{ia}{}^i{}_j = 0 \, . \tag{49}$$

Eqs. (48) and (49) imply the vanishing of the Sp(1) connection so that finally we end up with a hyper-Kahler manifold.

A characteristic of these manifolds is the existence of an almost quaternionic structure

$$(\Omega^{ij})^x{}_y = f^{x(i}_a \, f^{j)a}_y$$

which are covariantly constant

$$\nabla_z (\Omega^{ij})^x{}_y = 0 \tag{50}$$

(because $\omega_x{}^i{}_j = 0$).

The effect of N=2 supergravity coupled to the hypermultiplet is to resuscitate $\omega_x{}^i{}_j$. The holonomy group is then a subgroup of the full Sp(1) x Sp(n/4), and the spaces are the non-compact quaternionic manifolds[12]].

The coupling of the hypermultiplet to N=2 Yang-Mills fields leads us to the investigation of the symmetries of $\mathcal{M}$ which commute with supersymmetry. A symmetry transformation

$$\delta \phi^x = \xi^x_{(r)} \, c^{(r)} \tag{51}$$

is compatible with the superfield constraint (45) if the Killing vectors of G satisfy[5]

$$\xi_{(r)}{}^{a(i}{}_{;j}{}^{k)}{}_{b} = 0 \qquad (52)$$

where $\dot{j}$ indicates covariant differentiation with respect to the hyper-Kahler spin correction.

In the flat case where

$$\xi^{ia} = c^{ia} + L^a{}_b \phi^{ib} + L^i{}_j \phi^{ja} \qquad (53)$$

eq.(52) implies $L^i{}_j = 0$.

Thus the subgroup G_S of G, which commutes with supersymmetry should be contained in the semidirect product of Sp(n/4) and translations of $\mathcal{M}$.

The transformation of the spinors

$$\delta\lambda^a_\alpha = L_{(r)}{}^a{}_b\, c^{(r)} \lambda^b_\alpha \qquad (54)$$

$$L_{(r)}{}^a{}_b = \frac{1}{2}\left[\xi_{(r)}{}^{ai}{}_{;ib} - 2\omega_x{}^a{}_b\, \xi^x_{(r)} \right]$$

implies that only the linearly realized subgroup of G_S ($\subset$ Sp(n/4)) acts on the fermions. Eq.(52) can be solved in terms of a triplet function $\mathcal{X}^{ij}_{(r)}$ of ϕ^x

$$\xi^{ai}_{(r)} = -\frac{2i}{3}\, \mathcal{X}^{ij}_{(r)}{}_{;j}{}^{a} \qquad (55)$$

which satisfy

$$\mathcal{X}^{(ij,ak)} = 0 \; . \qquad (56)$$

The existence of $\mathcal{X}^{ij}_{(r)}$ is required from Noether's theorem and supersymmetry. $\mathcal{X}^{ij}_{(r)}$ is the first component of the linear multiplet[13] ($D^{(i}_\alpha \mathcal{X}^{jk)}_{(r)} = 0$) containing the conserved current associated with the symmetry.

Conversely, assuming that $\mathcal{X}^{ij}_{(r)}$ transforms homogeneously under G_S

$$\mathcal{X}^{ij}_{(r),x}\, \xi^x_{(s)} = f_{rs}{}^t\, \mathcal{X}^{ij}_{(t)} \qquad (57)$$

enables us to express $\mathcal{X}^{ij}$ in terms of the Killing vectors of G_S for G_S semisimple

$$\mathcal{X}^{ij}_{(t)} = i\,c\, f_{rst}\, \xi^{(r)(i}{}_{a}\, \xi^{(s)j)a} \quad , \quad c = \frac{\dim G_s}{f^{rst} f_{rst}} \; . \qquad (58)$$

The gauging of a subgroup, say H, of G_S is achieved by the minimal prescription. The constraint (45) is replaced by

$$f^{(ia}{}_x \left(D^{j)}_\alpha \phi^x + \mathcal{A}^{j)(\hat{r})}_\alpha\, \xi^x_{(\hat{r})} \right) = 0 \qquad (59)$$

where $\mathcal{A}_\alpha^{i(\hat{r})}$ is the spinor gauge potential in the adjoint representation of H. Following this procedure we can gauge any symmetry, linear or non linear, of the scalar manifold, provided it is compatible with supersymmetry.

Once the Lagrangian and transformation laws are obtained in d=6, we can perform non trivial dimensional reduction to d=5 and 4. In doing so we may restrict ourselves to symmetries which have not been gauged i.e.

$$\partial_d \phi^x + \mathcal{A}_d^{(\hat{r})} \xi^x_{(\hat{r})} = \xi^x_{(\hat{s})} m^{\hat{s}} + \mathcal{A}_d^{(\hat{r})} \xi^x_{(\hat{r})} \equiv \xi^x \cdot (m + \mathcal{A}_d) \tag{60}$$

where $\mathcal{A}_d$ becomes a scalar in d-1 dimensions.

As we have shown a non trivial reduction gives rise to potential terms. Because these are related to non-vanishing central charges, and since there is no potential term in the action in d=6, this method generates all the allowed potential terms for N=2 matter in d=5 and 4.

The result in d=4 is

$$\begin{aligned} V = \; & \tfrac{1}{2} \, g_{xy} \, \xi^x \cdot (m + \mathcal{A}) \; \xi^y \cdot (m + \mathcal{A}^*) \\ & + \tfrac{1}{8} \left[\kappa i (\mathcal{A} - \mathcal{A}^*) - i (\mathcal{A} \times \mathcal{A}^*) \right]^2 \\ & + \tfrac{1}{2} \, c^2 \, \Omega^{ij}{}_{xy} \, \Omega_{wz\,ij} \, (\xi^x \times \xi^y)_H \cdot (\xi^w \times \xi^z)_H \\ & - \; i c \, \Omega^{ij}{}_{xy} \, c_{ij} \cdot (\xi^x \times \xi^y)_H \end{aligned} \tag{61}$$

The first two terms arise from the reduction. The third, quartic in the Killing vectors of H, is the square of the auxiliary fields $x^{ij} x_{ij}$ as given by eq.(58). The last term is the Fayet-Iliopoulos term, where the constants c_{ij} can be non zero only for U(1) factors of H.

In our geometrical approach to interacting N=2 supersymmetric theories we have tried to be as general as possible, letting only supersymmetry impose restrictions on the geometry.

Acknowledgements
I am grateful to P.K. Townsend, M. Gunaydin and P.S. Howe for helpful discussions.

References

[1] G. Sierra and P.K. Townsend, Proc. 19th Karpacz Winter School of Theoretical Physics, 1983.

[2] E. Witten, "Global aspects of current algebra", Princeton preprint 1983.

[3] J. Scherk and J. Schwarz, Phys. Lett. 82B (1979) 60 ; Nucl. Phys. B153 (1979) 61.

[4] J. Bagger and E. Witten, Phys. Lett. 118B (1982) 103.

[5] G. Sierra and P.K. Townsend, LPTENS 83/21

[6] E. Cremmer, "Superspace and Supergravity", Proc.1980 Nuffield Workshop ;
E. Cremmer and B. Julia, Nucl. Phys. B159 (1979) 141.

[7] B. Zumino, Phys. Lett. 87B (1979) 203.

[8] E. Cremmer, B. Julia, J. Scherk, S. Ferrara, L. Girardello and P. van Nieuwenhuizen, Nucl. Phys. B147 (1979) 105 ;
E. Witten and J. Bagger, Phys. Lett. 115B (1982) 202.

[9] L. Alvarez-Gaumé and D.Z. Freedman, Comm. Math. Phys. 80 (1981) 443.

[10] P. Breitenlohner and Kabelschacht, Nucl. phys. B148 (1979) 96 ;
J. Koller, preprint CALT-68-975 (1983)
P.S. Howe, G. Sierra and P.K. Townsend, Nucl. Phys. B221 (1983) 331.

[11] G. Sierra and P.K. Townsend, Phys. Lett. B124 (1983) 497.

[12] J. Bagger and E. Witten, Nucl. Phys. B222 (1983) 1.

[13] P.S. Howe, K.S. Stelle and P.K. Townsend, Nucl. Phys. B192 (1981) 332.

BLACK HOLES IN COMPACTIFIED SUPERGRAVITY*

F.A. Bais

Institute for Theoretical Physics
Princetonplein 5, P.O. Box 80.006
3508 TA Utrecht, The Netherlands

ABSTRACT

Some consequences of dimensional compactification for black holes are discussed in the context of d = 11 supergravity. These include novel features concerning the structure of singularities and horizons.

1. Dimensional Compactification

Recently, we are witness of a serious revival of the old Kaluza-Klein idea in an attempt to reconcile unification with extended supergravity[1)]. One never knows: today's fiction may be tommorrow's paradigm. So far most of these attempts have been dealing with what one might call global or rigid compactification, that is, "above" each point in four-dimensional spacetime, one has the same compact seven-dimensional manifold. This raises a question. Is there such a thing as local compactification? Is it for example, possible to have solutions to d = 11 supergravity where the coordinates y^{α} of the internal space do depend nontrivially on the spacetime coordinates x^{μ}? Here, two rather obvious possibilities of potential interest come to mind:

i) Cosmological models, where the spontaneous compactification would presumably correspond to some dramatic event in the very early universe. This idea has been considered by Freund and others[2)].

ii) Black holes, where one naturally expects that the eleven-dimensional character of the theory plays an essential role near the

*Status report on work in collaboration with P. van Baal and P. van Nieuwenhuizen.

core of the black hole, typically at a scale where the curvature of the physical space becomes comparable to that of the internal space. Here, one may phantasize about novel features which might occur due to the higher dimensionality of the full space. For example, the singularity and horizon structure may be affected in an essential way (may be singularities can be avoided), the gravitational coupling constant may acquire x^{μ} coordinate dependence, etc.

In what follows, I will report on work in collaboration with P. van Baal and P. van Nieuwenhuizen[3,4)], which led to the conclusion that black holes in compactified supergravity do indeed exhibit novel features. It makes me think of black holes as the "deep throat" revealing the secrets of the internal space.

There are basically two different ways in which one may achieve a nontrivial coupling between spacetime M_4 and the internal space M_7. One is by introducing off diagonal elements in the eleven-dimensional metric which carry both a spacetime and an internal index. This possibility is naturally present in certain cases where one imposes a group G of isometries on the complete metric, as was pointed out by Manton and others[5)]. A second possibility, which is actually the one we pursue here, is to achieve the interaction between the two spaces by the stress-energy tensor of some matter field. We prefer this possibility as it appears to be simpler and moreover because in d = 11 supergravity, a bosonic matter field (namely the totally antisymmetric three index photon field A_{MNP}) is present in the theory. Whereas in the first possibility the coupling is intrinsically gravitational (i.e., through the left hand side of the Einstein equations), the second mechanism is more indirect (i.e. through the right hand side).

2. d = 11 Supergravity

The bosonic sector of d = 11 supergravity features besides the metric (vielbein) a three index photon field. The coupled field equations take the following form:

$$R_{MN} - \frac{1}{2} g_{MN} R = -96 \left(F_{MP_1P_2P_3} F_N^{\ P_1P_2P_3} - \frac{1}{8} g_{MN} F^2 \right) \tag{2.1}$$

$$D_M F^{MNPQ} = \frac{i\sqrt{2}}{48} \epsilon^{NPQP_1 \cdots P_4 Q_1 \cdots Q_4} F_{P_1 \cdots P_4} F_{Q_1 \cdots Q_4} \ . \tag{2.2}$$

We will adopt the following conventions: Capital latin indices, M,N,... run from 1 to 11, $\mu,\nu\ldots$ and m,n,... denote 4-dimensional curved and flat indices respectively. $\alpha,\beta,\ldots$ and a,b,... are reserved for the seven-dimensional curved and flat indices. The photon curl has strength unity and $R_{MN} = \partial_N \Gamma^\Lambda_{M\Lambda} - \partial_\Lambda \Gamma^\Lambda_{MN} + \ldots$ Dirac matrices satisfy $\{\Gamma_M, \Gamma_N\} = 2\delta_{MN}$.

By now a large number of solutions to the system (1) and (2) are known characterized by the internal symmetry group (i.e. the isometry group of M_7) and the number of surviving supersymmetries. Only two of them enter our considerations, we briefly discuss them in turn.

i) The Freund-Rabin solution[6], where

$$F_{mnrs} = i b \epsilon_{mnrs} \ ; \quad F_{abcd} = 0 \tag{2.3}$$

and $M_7 \simeq S^7$, i.e. a round seven sphere of radius R which through the equations is determined in terms of b as

$$b^2 = \frac{1}{32 R^2} \ . \tag{2.4}$$

The effective four-dimensional theory is the N = 8 theory with SO_8 gauge invariance[7].

ii) The Englert solution[8], where

$$F_{mnrs} = ib\epsilon_{mnrs} \, , \; F_{abcd} = cR^{-4}\bar{\eta}\Gamma_{abcd}\eta \; . \tag{2.5}$$

Here η is a covariantly constant (eight component) spinor on S^7 which satisfies the equation

$$D_\alpha \eta = \frac{i}{2}\Gamma_\alpha \eta \; . \tag{2.6}$$

In this case, the constants b and c are related to the radius R of S^7 as follows,

$$b = \frac{1}{6\sqrt{2}R} \, , \qquad c = \frac{1}{12\sqrt{2}}R^3 \; . \tag{2.7}$$

The effective four-dimensional theory has no supersymmetry left and has a residual SO_7 gauge invariance.

3. Black Hole Ansätze

We are interested in black hole solutions in d = 11 supergravity, which for large values of r, the radial coordinate of M_4, approach a compactified solution, specifically the Freund-Rubin- or Englert-solution. The question is then to what extent the extra dimensions may alter the interior aspects of a solution which asymptotically is just like an "ordinary" black hole. It does not take a great deal of imagination to write down the following ansatz for the line element

$$ds^2 = -B(r)dt^2 + A(r)dr^2 + r^2 d\Omega_2^2 + R^2(r)d\Omega_7^2 \tag{3.1}$$
$$(d\Omega_n)^2 = d\psi_n^2 + \sin^2\psi_n \, d\Omega_{n-1}^2 \; .$$

It is the direct sum of a Schwarzschild type line element for M_4 and a seven sphere with r dependent radius R. The symmetry group is obviously $R_1 \times SO_3 \times SO_8$.

Before giving the ansatz for the photon field, we observe that because D_α in (2.6) does not act on the variable r a covariant constant spinor still exists. It is convenient to rescale the vielbein on S^7 by introducing

$$\hat{e}_\alpha^a(y) = R^{-1}(r) e_\alpha^a(r,y). \tag{3.2a}$$

Now the ansatz for the photon simply reads

$$A_{\alpha\beta\gamma} = i c(r) \bar{\eta} \hat{\Gamma}_{\alpha\beta\gamma} \eta = i c(r) R^{-1}(r) \bar{\eta} \Gamma_{\alpha\beta\gamma} \eta \tag{3.2b}$$

so that

$$F_{r\alpha\beta\gamma} = \frac{1}{4} \partial_r A_{\alpha\beta\gamma} = \frac{i}{4} c'(r) \bar{\eta} \hat{\Gamma}_{\alpha\beta\gamma} \eta \tag{3.2c}$$

$$F_{\alpha\beta\gamma\delta} = c(r) \bar{\eta} \hat{\Gamma}_{\alpha\beta\gamma\delta} \eta .$$

At this point, we should make the observation that as long as we take b, c and R constant, any four-dimensional Einstein space (with cosmological constant) provides a solution, in particular we may take the usual Schwarzschild solution. We will denote the corresponding solution of the eleven-dimensional theory as Freund-Rubin-Schwarzschild and Englert-Schwarzschild solutions respectively.

4. The Radial Equations

Upon substitution of the ansätze (3.1-2) into the field equations, a coupled system of purely radial equations will result. The details of this reduction can be found in reference 3. Here we only quote the final form obtained after some ingeneous scrambling of the Einstein-Maxwell equations and the Bianchi identity. The important thing is that A can be written entirely in terms of R, b and c as follows

$$A(r) = \frac{r[r(\ln R)']' + 12(rc')^2[7r(\ln R)' + 2]R^{-6}}{r(\ln R)(r^2\omega_\Theta - 1) + r^2(6R^{-2} - \omega_\alpha)} \tag{4.1a}$$

with

$$\omega_\Theta = -192(2b^2 + 7c^2R^{-8}), \quad \omega_\alpha = 192(b^2 + 5c^2R^{-8}) . \tag{4.1b}$$

This allows one to eliminate A and B altogether resulting in a coupled system for R, b and c:

$$\begin{aligned} 2r(\ln A)' = {} & R^{-8}[r^2(R^7)']'[1 + \tfrac{r}{2}(\ln R^7)']^{-1} \\ & + 2 - 2A(1 - r^2\omega_\Theta) + 84(rR^{-3}c')^2 \\ & \times [1 - r(\ln R^7)'][1 + \tfrac{1}{2}(\ln R^7)'r]^{-1} \end{aligned} \tag{4.2}$$

$$\begin{aligned} c'' + r^{-1}c\,[1 + A - 6rR'R^{-1} - r^2A\omega_\Theta + 84(rc')^2R^{-6}] & \\ + 16cAR^{-2}(6\sqrt{2}\,bR - 1) = 0 & \end{aligned} \tag{4.3}$$

$$(bR^7)' = -21\sqrt{2}(c^2)' . \tag{4.4}$$

Substituting (4.1) into (4.2), one arrives at a third order equation for R. The Maxwell equation (4.4) is integrable and allows one to eliminate b in terms of R and c (and some integration constant). So the problem has been reduced to a coupled system for R and c only. This has some advantages. One should realize that we do not in general have the relation AB = 1, which usually makes life much easier as far as constructing exact solutions is concerned. In fact, to determine the properties of the solutions, we had to subject ourselves to the use of a computer. Solving Einstein's equations numerically is usually a delicate business because of the appearance of horizons etc. It is therefore of importance to isolate suitable variables which are well behaved for the whole range of r (excluding maybe the origin). R is such a variable, in fact one expects on physical grounds that if R would become singular at some finite value of r, this would correspond to a physical singularity making the solution rather uninteresting.

Let us now turn to the solutions obtained so far[3,4]. We distinguish three cases. In Case I we set b = c = o, i.e. we look for real eleven-dimensional black holes without compactification. In Case II, we still set c = o , but take b ≠ o, i.e. solutions without torsion, where asymptotically we have the Freund-Rubin-Schwarzschild solution. Case III represents the general case where both b ≠ o and c ≠ o. There is a suggestive analogy with the 't Hooft-Polyakov monopole solution in the Yang-Mills-Higgs system, which leads one to believe in the existence of a completely regular solution.

5. Case I: Eleven-Dimensional Black Holes Without Compactification

We set b = c = o in which case the system (4.1-2) simply reduces to

$$A = [r(\ln R)']' \, [-r(\ln R)' + 6r^2 R^{-2}]^{-1} \tag{5.1a}$$

$$2r\frac{A'}{A} = 2(1-A) + R^{-8}[r^2 R (R^7)']' \left[1 + \frac{r}{2}(\ln R^7)'\right]^{-1} . \tag{5.1b}$$

In order to find an exact solution, it suffices to make the educated guess $R(r) = \alpha r$, substitution in (5.1a) yields the condition $\alpha^2 = 6$, whereas (5.1b) takes the form

$$r A' A^{-1} = 8 - \Lambda \tag{5.2}$$

with the exact solution:

$$R = r\sqrt{6}, \quad A = 8(1 - mr^{-8})^{-1}, \quad AB = 1. \tag{5.3}$$

The following observations can be made:

i) The solution (5.3) exhibits a horizon with an $S^2 \times S^7$ topology;

ii) The space is not asymptotically flat because of the factor 8 in the numerator of A. This in contrast with the standard black hole in eleven dimensions with an S^9 horizon obtained from the line element

$$ds^2 = -Bdt^2 + Adr^2 + r^2 d\Omega_9^2 \tag{5.4}$$

with solution

$$A = (1 - 2mr^{-8})^{-1}, \quad AB = 1. \tag{5.5}$$

iii) The same recipe applies to black holes with topologies like $S^3 \times S^6$, $S^4 \times S^5$, etc.

6. Case II: The Compactified Black Hole Without Torsion

We set c = o but b ≠ o. We fix the scale by putting $R(\infty) = 1$, from (4.4) we know that bR^7 = constant, whereas the asymptotic value of b is given by (2.4) as $b(\infty) = 1/32$ so that

$$b^2 = \frac{1}{32} R^{-14} \tag{6.1}$$

The equations (4.1-2) take the form

$$A(r) = r\,[r(\ln R)']'\,[-r(\ln R)'(1 + 12r^2R^{-4}) + 6r^2(R^{-2} - R^{-14})]^{-1} \tag{6.2a}$$

and

$$2\frac{A'}{A}=7\frac{R''}{R}\left[\frac{1}{r}+\frac{7R'}{2R}\right]^{-1}+\frac{2}{r}(1-A)+14\frac{R'}{R}-24\frac{rA}{R^{14}} \quad . \tag{6.2b}$$

An expansion around the asymptotic behaviour at $r = \infty$ shows that the solutions are characterized by two free parameters, a mass m and some effective charge q, as follows:

$$R=1+qr^{-6}\left(1-\frac{15}{44}r^{-2}+\frac{m}{2}r^{-3}+\ldots\right) \tag{6.3a}$$

$$A=\left(1-2mr^{-1}+4r^{2}\right)^{-1}\left(1-49qr^{-6}\{1+O(r^{-2})\}\right) \tag{6.3b}$$

$$B=\left(1-2mr^{-1}+4r^{2}\right)\left(1-\frac{7}{4}qr^{-8}\{1+O(r^{-2})\}\right) \quad . \tag{6.3c}$$

Note that the term $4r^2$ in the prefactor of A and B comes from the cosmological constant which is induced by the nonvanishing asymptotic value of b. It is in fact possible to have a vanishing physical cosmological constant[4)], by adding a (fine tuned) cosmological term Λ^* to the eleven-dimensional theory. This term breaks supersymmetry though, and furthermore the asymptotics (6.3) is altered in an essential way: instead of power corrections, one gets exponential behaviour.

In figure 1 we exhibit the solution with $m = 100$ and $q = 500$, which was calculated numerically.

The behaviour of this particular solution is representative for any solution with $q > 0$. The solution is entirely regular for finite values of r. For $r \to \infty$, the singular behaviour of B is entirely due to the $4r^2$ term of the cosmological constant. This singularity is harmless as it reflects the confining nature of the effective potential for a massive test particle. It would take an infinite energy to ever reach the $r = \infty$ singularity.

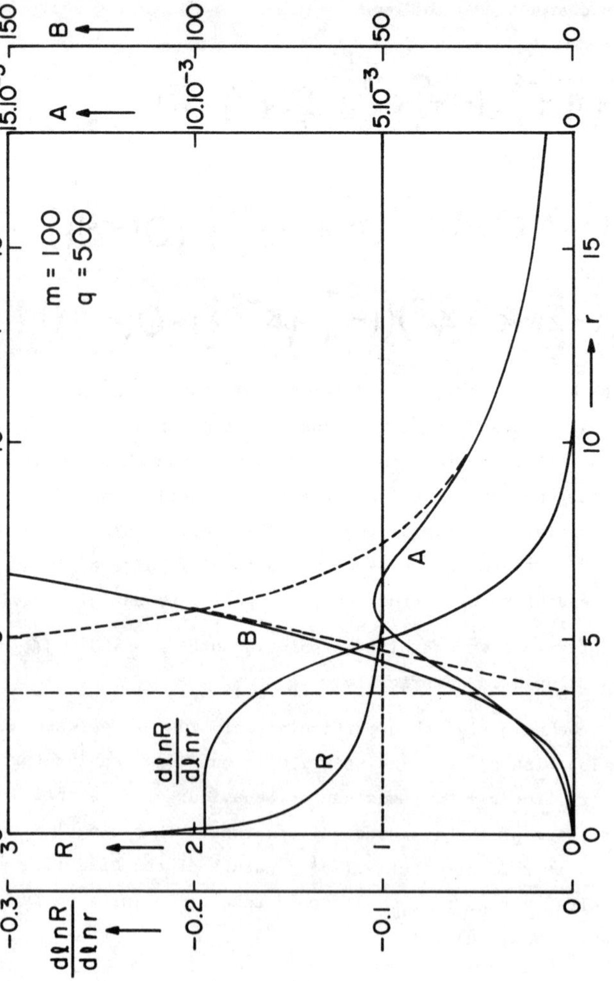
m = 100
q = 500
r
A
B
R
dℓnR/dℓnr
15.10⁻³
10.10⁻³
5.10⁻³
150
100
50
-0.3
-0.2
-0.1
0
1
2
3
5
10
15

Figure 1

Let us now turn to the behavior as $r \to o$. Three remarkable things happen. Firstly, the radius R of the internal space tends to infinity. Secondly, because $B(o) = o$, we have an horizon at $r = o$. Finally, there is also a physical singularity at the origin because $A(o) = o$. The coincidence of these various singularities clearly distinguishes this solution from the conventional Schwarzschild or Reissner-Nordstrøm solutions, and demands a detailed analysis of the causal structure around $r = o$. These causal features we wish to discuss are all contained in the Penrose diagram, which was constructed in reference 4 and is exhibited in figure 2.

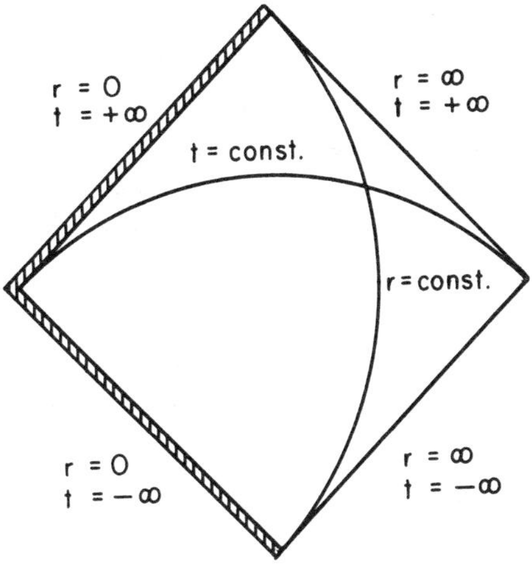

Figure 2

From the picture, it is clear that the spacetime we are considering does not have any causal inconsistencies. The singularity at r = o corresponds apparently to a light like singularity which nicely complements the well-known space- and timelike singularities which occur in the Schwarzschild- and Reissner-Nordstrøm solutions respectively. We are apparently dealing with a "tightly dressed" singularity. It should be noted that because horizon and singularity coincide, it is not possible to analytically continue beyond the horizon. It is impossible to connect other universes to the singularity in contrast with the ordinary black hole solutions. This fact has another important consequence related with the Hawking temperature. The Hawking temperature is directly related to the occurrence of an horizon. However, to determine this temperature, it is crucial that there exists a coordinate system which is well behaved over the horizon. Continuing these coordinates to imaginary time, the Hawking temperature is found to equal the inverse period of the Euclidean coordinates. In the case at hand, there is no coordinate system which is regular over the horizon, the t-dependence can be scaled arbitrarily and the Hawking temperature is undetermined. Quantum gravity becomes indispensible at this point.

7. Case III. The Compactified Black Hole with Torsion

We consider the general case where we set both b $\neq$ o and c $\neq$ o, i.e., we look for solutions which asymptotically tend to the Englert solution (2.5-7). The analysis of this case has not yet been completed and we will only present a speculation based on an amusing analogy with the 't Hooft-Polyakov monopole solution in the Yang-Mills-Higgs system. The analogy suggests the existence of a completely regular soliton like solution to compactified supergravity.

We first recall some basic facts concerning the monopole case and then make the comparison. The SO_3 Yang-Mills model with a triplet of gauge potentials A_μ^a and scalars ϕ^a has the following monopole solutions.

i) The Wu-Yang[9] solution where $\phi^a = o$ and $A_i^a = \varepsilon_{aib} r^b/r^2$. It is a solution to the pure Yang-Mills equations which is point-singular (and in fact unstable).

ii) The point-singular solution where $\phi^a = r^a/r$ and $A_i^a = \varepsilon_{aib} r^b/r^2$. In this case ϕ is gauge covariantly constant on the two sphere. To make the similarity more striking, one could write $\phi^a = c\eta^\dagger\tau^a\eta$ with τ^a the Pauli matrices and η an SU_2 spinor, where

$$\eta = g^{-1}\eta_0 , \quad g^{-1} = \begin{bmatrix} \cos\Theta/2 & -\sin\frac{\Theta}{2}e^{-i\phi} \\ \sin\frac{\Theta}{2}e^{i\phi} & \cos\frac{\Theta}{2} \end{bmatrix} , \quad \eta_0 = \begin{pmatrix} 1 \\ 0 \end{pmatrix} . \tag{7.1}$$

Because now $d\eta = g^{-1}dg\eta$, the spinor is SU_2 covariantly constant.

iii) The regular 't Hooft-Polyakov monopole[10] is just $\phi^a = c(r)\eta^\dagger\tau^a\eta$ and $A_i^a = w(r)\ \varepsilon_{iab} r^b/r^2$ when the functions c and w smoothly interpolate between $c(\infty) = w(\infty) = o$ and $c(o) = w(o) = 1$.

The analogy we have in mind is obvious at this point. The field $F_{\alpha\beta\gamma\delta}$ in (5) can be interpreted as a Higgs field which breaks the SO_8 symmetry down to SO_7[11] (this is not the isotropy group of the compact S^7 though). We think of the Freund-Rubin-Schwarzschild solution (c = o) as the Wu-Yang solution. The Englert-Schwarzschild solution (c = const.) is the analogue of the point-singular solution. (This raises the question of multiply "charged" Englert solutions.) We envisage the analogue of the regular monopole as a soliton in d = 11 supergravity with $c(\infty)$ and $R(\infty)$ finite, but with $c(o) = 0$. If in addition R(o) and b(o) would be also finite and nonvanishing, one would have a solution which interpolates between an Englert solution at $r = \infty$ and a Freund-Rubin solution at r = o.

It is well-known that the existence of the 't Hooft-Polyakov monopoles is intimately related to the topology of the Higgs field. In fact the nonvanishing value of ϕ^a in ii) breaks the symmetry from $G = SO_3$ to $H = SO_2$. The Higgs field on the two sphere S_R^2 at spatial infinity ($r \to \infty$) maps a point (θ,ϕ) into the coset space $G/H = SO_3/SO_2 \simeq S^2_\phi$. To see this recall that the covariantly constant Higgs field (or the covariantly constant spinor in that matter) may be written as

$$\phi(\theta,\varphi) = g(\theta,\varphi)\phi_o \tag{7.2}$$

where $g \; G$ and ϕ_o is really constant (e.g. $\phi^a_o = \delta^a_3$). We can decompose g always as a product of a subgroup element h and a coset element k as

$$g = kh$$

so that

$$\phi(\theta,\varphi) = k(\theta,\varphi)\phi_o \quad . \tag{7.3}$$

Therefore one has that the boundary conditions on the Higgs field fall into topologically distinct homotopy classes labeled by the degree (winding number) of the mapping

$$k: S^2_R \to S^2_\phi \tag{7.4}$$

i.e. by elements of the second homotopy group $\pi_2(G/H) \simeq \pi_2(S^2_\phi) = Z$ (the group of integers under addition). A similar situation is encountered in the Englert solution where we have that $F_{\alpha\beta\gamma\delta}$ depends on the coordinates of the compactified space S^7_R (to be compared with S^2_R) which provides us with a mapping into the coset space $G/H \simeq SO_8/SO_7 = S^7_F$. The topological quantum number in this case is thus associated with an element of the homotopy group

$$\pi_7(S^7_F) \simeq Z \quad . \tag{7.5}$$

We should emphasize that S_R^7 and S_F^7 are distinct seven spheres related to two nonequivalent SO_7 subgroups of SO_8. At the other hand, since both are subspaces of the same SO_8, it may well be that the winding number is fixed. This is also suggested by the fact that the covariantly constant spinor used by Englert is in fact a unique solution up to a sign in equation (6) and some global SO_8 rotation. This is different from the monopole case where the boundary condition on the Higgs field can generate any winding number. However, the situation we discuss is very similar to the monopole case if one in addition would insist on spherical symmetry under the combined angular momentum $\vec{J} = \vec{L} + \vec{T}$ as in (7.1), in which case one also has only the solutions with winding number plus or minus unity.

One may pose the following question. As the photon field we are dealing with here has nonlinear self interactions (as is evident from (2.2)), in spite of being abelian, one may wonder whether it has soliton-like solutions in a fixed gravitational background. Assuming this background field to correspond to the Englert-Schwarzschild solution (m = o), one obtains from (4.3) the following equation

$$\ddot{c} = f(r)\dot{c} + \alpha c\left(c^2 - c_\infty^2\right) \tag{7.6}$$

where α = constant and f(r) is finite for r > o. The variable t is chosen such that it satisfies

$$\frac{dt}{dr} = -\left(1 + 10r^2/3\right)^{1/2} \tag{7.7}$$

Does (7.6) admit a solution with $c(t = -\infty) = c_\infty$ and $c(t = +\infty) = o$? This question is easily answered using a mechanical analogue. The equation describes the motion of a point particle (coordinate c = c(t)) in a one-dimensional potential $v(c) = -\alpha/4(c^2 - c_\infty^2)^2$. In addition, it experiences the utopian property of negative friction. It is exactly this "wrong" sign which causes that the desired solution does not exist. This is not so surprising in view of our analogy where we want the gravitational fields A, B and R to play the same role as the gauge fields in the monopole case, and these are indispensible for obtaining the regular 't Hooft-Polyakov solution. It is amusing to note that the same equation (7.6) arises for the Higgs field if one fixes the gauge field to be the Wu-Yang solution.

At present we are involved in a numerical enterprise to search for a regular solution. As was shown in reference 3, this necessitates a 3 parameter search. So far we have been able to get rid of the horizon for any value of m, but have not yet been able to establish the existence of a soliton. We hope to settle the question in the near future.

Acknowledgements: I thank the organizers of the Workshop for their hospitality. This work has been supported by the Dutch National Science Foundation (FOM and ZWO).

References

1. P.G.O. Freund and M.A. Rubin, Phys. Lett. 97B, 233 (1980).
   M.J. Duff and C.N. Pope, CERN-TH 3451 (1982).
   F. Englert, talk presented at this meeting.*
2. P.G.O. Freund, Phys. Lett. 120B, 335 (1983); Nucl. Phys. B209, 146 (1982); I.B. Benn and R.W. Tucker, Phys. Lett. 25B, 133 (1983).
3. P. van Baal, F.A. Bais and P. van Nieuwenhuizen, Utrecht preprint (1983).
4. P. van Baal and F.A. Bais, Utrecht preprint (1983).
5. N. Manton, ITP-Santa Barbara preprint NSF-ITP-83-04 (1983); R. Cocquereaux and A. Jadczyck, CERN-TH 3483 (1983).
6. P.G.O. Freund and M.A. Rubin, Phys. Lett. 97B, 233 (1980).
7. B. de Wit and H. Nicolai, Phys. Lett. 108B, 285 (1982); Nucl. Phys. B208, 323 (1982).

---

* No written version of F.Englert's talk was received. (Editors)

7. (continued) M.J. Duff, CERN-TH 3451 (1982).

8. F. Englert, Phys. Lett. 119B, 339 (1982).

9. T.T. Wu and C.N. Yang, Properties of Matter under Unusual Conditions, (H. Mark and S. Fernback, Interscience, New York, 1969).

10. G. 't Hooft, Nucl. Phys. B79, 276 (1974).
A.M. Polyakov, JETP Lett. 119B, 339 (1982).

11. N. Warner, Caltech preprint 68-1008 (1983);
B. de Wit and H. Nicolai, Nikhef preprint (1983).

"LOW ENERGY" SUPERSYMMETRY AND SUPERGRAVITY

Riccardo Barbieri
Dipartimento di Fisica
Università di Pisa
I.N.F.N., Sezione di Pisa

1. Introduction and minimal models

Supersymmetry is the only way that we know of for producing a naturally light fundamental scalar Higgs field. It looks therefore interesting to consider a possible extension of the standard model to a supersymmetric theory. Unlike the case in which one tries to relate the Fermi scale to a new strong interaction, one may insist here on a purely perturbative theory, which means in turn more reliable calculation versus a lower potential predictive power. On the other hand, the focus on the natural Fermi scale problem is certainly not the only reason to consider a realistic supersymmetric model. In many respects one may think that supersymmetry is the remaining symmetry to be explored in a field theory description of fundamental particle interactions.

The possibility of producing a naturally light scalar Higgs field φ rests on the validity of the sum rule

$$\sum_{J=0,1/2,1} (2J+1)\,(-)^{2J}\, M^2_J(\varphi) \equiv \mathrm{STr}M^2_J = 0 \qquad (1)$$

among the masses $M_J(\varphi)$ of the particles of spin J coupled to φ, as functions of φ itself. Eq. (1) holds automatically in a large class of supersymmetric models, due to the cancellation occurring between fermions and bosons of each supermultiplet.

This same equation, however, is also the source of most problems encountred in discussing a realistic model since it gives, more or less directly, a serious constraint on the actual masses $M_J(\langle\varphi\rangle)$. In minimal extensions of the standard model one would obtain at least an unacceptably light scalar quark partner. This is a well known difficulty first pointed out by Fayet.

To overcome it another unique property of supersymmetric theories is of help. Models can be constructed where a naturally tiny trasmission of the supersymmetry breaking takes place to a full sector of the particle spectrum. This occurs because of the possibility of decoupling different physical scales in a suitably constructed weak coupling supersymmetric lagrangian. Typical models include the "light" Fermi scale (M_w), the supersymmetry breaking scale $M_s \gg M_w$ and a third "heavy" scale $M = O(M_s^2/M_w)$.

The real interest of the situation lies in the fact that it can be nicely implemented in supergravity with M identified the Planck scale M_P. The most economic scheme[1,2] uses, other than the "observable" light supermultiplets, only the Goldstino and its bosonic superpartner. If it would not be for the supersymmetrized gravity coupling, the Goldstino would not see at all the light sector. As a consequence , the physics we see, ignoring supergravity, would be entirely supersymmetric. On the other hand the smallness of the gravitational coupling at the scale of particle physics makes to survive only the effects involving the large supersymmetry breaking scale M_s. More precisely, M_s is determined to be

$$M_s^2 = O(M_W M_P) \simeq (10^{10-11}\,\mathrm{Gev})^2. \tag{2}$$

The otherwise massless Goldstino field, through the superHiggs effect,[3] gets eaten by the spin 3/2 gravitino, which acquires a mass

$$m_g = \frac{M_s^2}{\sqrt{6}\,M_P} = O(M_W). \tag{3}$$

In explicit models, the vacuum expectation value of the auxiliary component in the Goldstino superfield breaks supersymmetry at $M_s^2 = O(M_W M_P)$. After that, tourning off gravity ($M_P \to \infty$) in the classical Lagrangian,[1] gives in the low-energy effective Lagrangian for the light fields Y_a

$$\begin{aligned} L(Y_a) &= L(\text{glob supersym};\, Y_a, f) \\ &\quad - m_g^2\, Y_a{}^* Y_a - m_g\left((A-3) f(Y_a) + Y_a \frac{\partial f}{\partial Y_a} + \text{h.c.}\right). \end{aligned} \tag{4}$$

Here the first term is the globally supersymmetric Lagrangian depending on an arbitrary superpotential f. The explicit supersymmetry breaking terms, controlled by the gravitino mass m_g, are specified by f and the parameter A depending[4] on the particular way the x-field enters in the complete superpotential.

The gauge group that I shall consider is the standard one, G = SU(3) x SU(2) x U(1). Other than the standard fields (quarks, leptons and gauge bosons) all accompained by a related superpartner, the Y_a must include an enlarged supersymmetric Higgs sector to describe the breaking of $SU(2)_L \times U(1)_Y$: a pair of $SU(2)_L$ doublets H, H_c of opposite hypercharge to get the needed degrees of freedom and, minimally, an overall group singlet Y if one wants the breaking to occur at the tree level in a simple manner. There are objections to a light group singlet Y coming from the attempts to embed G in a grand unified group. For the time being I neglect them because: (i) to have a natural Fermi scale in the standard model is a useful achievement per se; (ii) in GUT's, letting aside the light singlet problem, even the problem of obtaining a natural splitting between the light Higgs doublets from their heavy colour triplet partners is still under discussion.

The appropriate model has finally emerged.[1,2] The superpotential $f(Y_a)$ will contain a term

$$Y(HH_c - \mu^2)$$

to drive the SU(2) x U(1) breaking, together with the supersymmetric Yukawa interactions. In fact $f(Y_a)$ may be the most general G-invariant polinomial of cubic degree at most, which respects Baryons and Lepton number (or a discrete symmetry which enforces their conservation).

Before going to a description of the particle spectrum, I want to emphasize a general aspect of the class of models that I am describing, i.e. the automatic conservation by strong interactions of flavour and CP, letting aside the θ-problem, as it is the case in the standard model. Otherwise gluino and scalar quark exchanges could lead to unacceptably large flavour changing neutral currents and/or to a dangerous neutron electric dipole moment. Also potentially large contributions to flavour and CP violations in the K_0-$\bar{K}_0$ system from supersymmetric weak interactions are kept under control.

2. Phenomenological implications

The class of models under consideration gives rise to a particle spectrum with the following salient features, independent from the details of the parameters in the superpotential f:

i) There is a charged w-ino, - a fermion partner of the W -, which is lighter than the W itself[5]. Its mass gets close to the W-mass as the gravitino mass decreases ($m_g \gtrsim$ 20 Gev), whereas it becomes light

$$m(\tilde{\omega}) = O(M_W^2/m_g) \qquad (5)$$

for an heavy gravitino ($m_g \gg M_W$). The other w-ino is heavier than the W.

ii) The masses of the two scalar particles associated with any (Dirac) quark and lepton, f, are related to m_f and m_g according to

$$\mu^2_{f,\pm} = m^2_g + m^2_f \pm 2\, b\, m_g m_f \tag{6}$$

b being a model dependent parameter of order unit. The charged scalar component in H, H_c, not eaten by the W, is substancially heavier than the W itself.

iii) In the neutral fermion sector there is always a Majorana spinor $\tilde{z}$ lighter than the Z, with a mass

$$m(\tilde{z}) = O(M_Z^2/m_g) \tag{7}$$

for $m_g \gg M_W$. The remaining part of the neutral fermion sector (3 more Majorana spinors a part from the photino) is more model dependent, as it is the spectrum of the neutral scalar H, H_c sector.

(iv) In the model there are light gluinos ($\tilde{g}$) and light photino ($\tilde{\gamma}$). At the tree level they are massless. They pick up radiative masses from the heaviest supermultiplets which have colour and charge respectively. At the one loop the top-scalar top finite contribution to $m(\tilde{g})$ never exceeds 1 Gev. Also at the one loop the photino, besides the mentioned top contribution, receives a mass from w-ino exchange at most as large as 100 ÷ 400 Mev. These masses appear too small to be consistent with observation and/or with cosmological limits (for a stable photino). At higher loops, though, logarithmic divergent contributions come in which must be cut off at $\Lambda \lesssim M_p$. Their size can be as large as the one loop finite contributions. This makes a precise prediction impossible but calls for masses in the few Gev range. Of course if many radiative contributions pile up from unknown heavy supermultiplets, the gluino and photino masses can get larger but still of order $(\alpha_s, \alpha) m_g$. On the other hand, I am reluctant to accept $m(\tilde{g})$, $m(\tilde{\gamma}) \simeq m_g$, since this is against the motivation for using the "minimal" Lagrangian, Eq.4. Hunting for "light" gluinos and for a "light" stable photino is an important tool in the search for supersymmetry evidence.

The properties of the spectrum that I have described, together with the coupling of the new particles as dictated by supersymmetry and gauge invariance, make definitely possible to test (disprove) the class of models discussed. This concrete possibility is perhaps offered already at the CERN $p\bar{p}$ collider[6]. In order to incorporate the Higgs picture in a natural framework, supersymmetry surrounds the Higgs scalar with a plethora of new particles which cannot be too elusive. Some of them might (should) already show up in the decay fragments of the newly found intermediate vector bosons.

3. References

1. R.Barbieri, S.Ferrara and C.Savoy, Phys. Lett. 119B, (1982) 343.
2. R.Arnowitt, A.Chamseddine and P.Nath, Phys.Rev.Lett. 49, (1982) 970.
3. S.Deser and B.Zumino, Phys. Rev. Lett. 38 (1977) 1433.
4. H.Nilles, M.Srednicki and D.Wyler, Phys. Lett 120B (1983) 346.
5. S.Weinberg, Phys.Rev. Lett. 50 (1983) 387.
6. For a complete description and list of references see R.Barbieri, Pisa Univ. preprint IFUP TH 84-14 (1983).

NUMERICAL SIMULATION ON THE ICL DISTRIBUTED ARRAY PROCESSOR

D.J. Wallace

Physics Department
The University
Mayfield Road
EDINBURGH EH9 3JZ
Scotland

ABSTRACT

We review some recent results of Monte Carlo simulation obtained on the Distributed Array Processor. After a brief description of its hardware and software features, we discuss Monte Carlo renormalisation group calculations for the three dimensional Ising model and SU(3) lattice gauge theory calculations for hadron masses in the quenched approximation.

1. Introduction

In this talk I shall review some of the calculations done recently on the ICL Distributed Array Processor (DAP) at Edinburgh, in collaboration principally with K.C. Bowler and G.S. Pawley. The DAP consists of a hard-wired square array of 2^{12} processing elements (PE's) which operate concurrently on their own data [1]. This architecture is readily exploited for most Monte Carlo simulations by discretising the physical system of interest and distributing the resulting lattice points over the array of PE's. The PE's can then provide up to 4096 updates, or time-steps, in parallel and for this kind of problem the machine's power approaches that of CRAY-1. The potential for future development of this architecture, with more sophisticated PE's and VLSI implementation, is high.

The main areas of physics which have been studied with the machine at Edinburgh are (i) molecular dynamics of plastic crystals and polar liquids (G.S. Pawley and collaborators [2]), (ii) Monte Carlo studies of phase transitions and (iii) lattice gauge theories. The three sections of this paper describe briefly the hardware and software features of the DAP and review some of the results obtained so far (June 1983) in (ii) and (iii). This paper is composed from previous workshop and conference talks [3] by K.C. Bowler and D.J.W; numerical results quoted remain preliminary.

2. Hardware and Software Features of the DAP

The basic hardware of the DAP consists of a 64×64 square array of processing elements with (optional) cyclic boundary conditions. Each PE has a 4k bit store (on the machine at Edinburgh) and a very simple one-bit processor with three registers. Certain instructions may be made conditional upon the setting of one of these, the activity or A-register. The manner in which this degree of local autonomy may be exploited is discussed below. Instructions are broadcast to the processor array by a master control unit (MCU) which also handles certain simple scalar functions such as control of DO loop variables in Fortran. The processors in the array obey each instruction simultaneously, acting on their local data. Thus the DAP is a single-instruction -stream, multiple-data-stream (SIMD) machine.

The DAP is constructed by allocating a processor to each memory chip of a standard 2Mbyte store module of an ICL 2900 series mainframe. There is thus no overhead associated with loading the DAP as there is with array or vector processors which are attached to a mainframe as back-end processors. Data in the DAP store may be processed either by the DAP or by the host computer. The other side of this coin is of course that the DAP is specific to the 2900 series of machines, although ICL have under development a mini-DAP with 32×32 16kbit PE's based on the PERQ mini.

In order to exploit the parallelism of the DAP, a dialect of Fortran known as DAP Fortran has been developed. The three basic types of entity in DAP Fortran are scalars, vectors and matrices. A scalar corresponds to an ordinary Fortran scalar whereas vectors and matrices are arrays of 64 and 64×64 items respectively, which may be treated as sets of scalars or, more efficiently, as complete arrays. Thus for example the addition of two 64×64 arrays can be accomplished by the following piece of code:

```
DIMENSION A(,),B(,),C(,)
A = B+C
```

with the calculation and assignment performed simultaneously on all elements.

Variables and constants may be of type REAL, of length 3 to 8 bytes, INTEGER, of length 1 to 8 bytes or LOGICAL, and are declared in a similar manner to normal Fortran. For example,

```
INTEGER*2 MBOND(,,4,18)
LOGICAL LMASK(,)
```

The second example declares a logical 64×64 array which can be thought of as residing in a single bit-plane of the DAP.

Two features of DAP Fortran are of special importance in exploiting the parallel architecture. The first such feature is logical masking: operations and assignments may be made conditional upon a logical mask defined over the 64×64 array of PE's, which in effect sets the A-register already referred to. Such masks can be defined using built-in logical functions available in DAP Fortran, or they may be dynamically generated within a program. Consider the following example:

```
LOGICAL LMASK(,)
LMASK=ALTR(1)
```

ALTR(N) is a built-in function which sets alternate blocks of N rows of the 64×64 array .FALSE.and .TRUE. respectively. Thus LMASK has the structure indicated in figure 1(a). We could instead set alternate

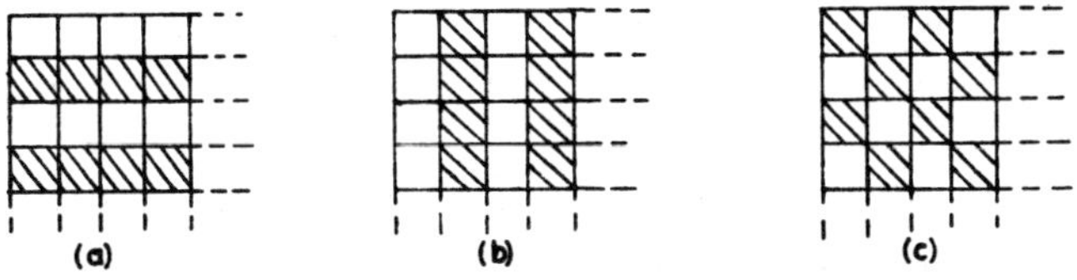

Figure 1. Examples of logical masks

columns of LMASK to .F.and .T. by means of the assignment LMASK=ALTC(1) as depicted in figure 1(b). More elaborate masks can be constructed using standard logical operators. Thus

```
LMASK=ALTR(1).LEQ.ALTC(1)
```

sets up the chequerboard pattern of figure 1(c). Assignments may now be made conditional upon such a mask. For example

```
REAL*4 A(,),B(,)
A(LMASK)=B
```

assigns the elements of B to the corresponding elements of A only at those PE's for which LMASK is .TRUE.

The second crucial feature is that of shifting: calculations at each PE may require information from a number of PE's, and this information is brought in by means of various shift operations. As an illustration consider

```
DIMENSION A(,),B(,),C(,)
A=B+SHWC(C,3)
```

The result is to assign to A at each PE the sum of the element of B which is stored at that PE and the element of C which is stored three sites away in an 'easterly' direction, with cyclic boundary conditions automatically imposed over the entire chain of 64 PE's in the east-west line. Similarly there are shifts east, north and south with either cyclic (SHEC,SHNC,SHSC) or planar (SHWP,SHEP,SHNP,SHSP) boundary conditions. It is also possible to use 64×64 arrays in long-vector mode: shifts along the 4096-vector are then accomplished by means of operations involving left or right shifts, namely SHLC,SHRC,SHLP,SHRP.

Finally in this section we discuss briefly some general features of the maps and algorithms which are exploited in these lattice simulations. The basic principle is to distribute the lattice sites of the physical system over the 4096 PE's. This mapping is obvious for some problems. For example the 64^3 Ising model is mapped on to 64 bit planes; for smaller lattices, an appropriate number of simulations are done simultaneously to increase statistics. There are many ways of mapping a four-dimensional hypercubic lattice on to the two-dimensional array of PE's of the DAP. Currently for SU(3) calculations we use an 8^4 lattice distributed as shown in figure 2; the 64×64 array is

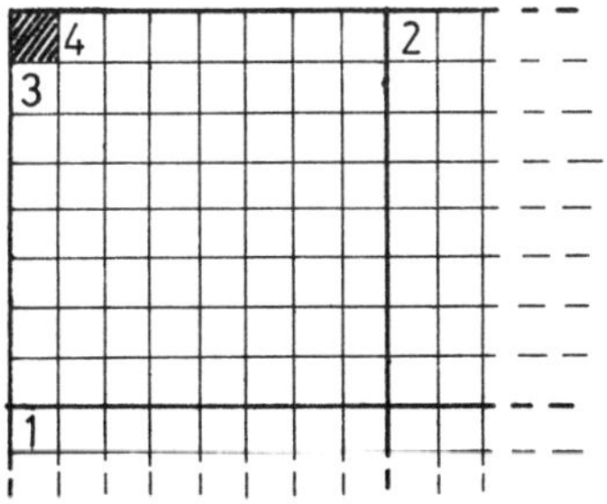

Figure 2. Nearest neighbours of a typical site in the mapping of an 8^4 lattice on to the DAP.

divided into 8×8 blocks of PE's; each PE corresponds to a single site of the 8^4 lattice, with two of the four dimensions mapped within blocks on the remaining 8×8 blocks.

With regard to algorithms, two points are worth making. First in order to ensure convergence to equilibrium, one should not attempt to update independently and concurrently two variables which interact with each other in the action. For the Ising model with nearest neighbour interactions this can be achieved by updating all the even sites of the lattice and then all the odd sites, in pairs of bit planes, with complete efficiency. In four-dimensional lattice gauge theories although it is possible to update in principle 1 in 4 link variables [4)], the data shifting required is not transparent and may be inefficient. Instead we update 1 link in 8, e.g. from the even sites in a given direction; the mask needed to label even and odd sites is 8×8 chequerboard of 8×8 chequerboards given by the DAP Fortran statement

```
LMASK = (ALTR(1).LEQ.ALTC(1)).LEQ.(ALTR(8).LEQ.ALTC(8))
```

In this case only $\frac{1}{2}\times 4096$ variables are being updated simultaneously; the machine is used efficiently by distributing over all the PE's the SU(3) matrix multiplies required to calculate the change in action. Second, the relaxation routine to calculate the quark propagator is implemented elegantly on an $8^3\times 16$ lattice by recursion on even → odd → even etc.

sites, since $\not{D}$ connects only nearest neighbours. Some details are given by K.C. Bowler (ref.3).

3. Monte Carlo Renormalisation Group for the 3-d Ising Model

The machine is particularly adapted for Ising model studies because of the powerful logical facilities provided in the DAP software. We [5] have studied the phase transition from the disordered phase at high temperature to the ferromagnetically ordered phase at low temperature, on lattices of 8^3, 16^3, 32^3 and 64^3 sites, with cyclic boundary conditions. We use the Monte Carlo renormalisation group (MCRG) approach to estimate the critical temperature and exponents for the phase transition. In the following sub-sections I review the basic ideas and equations of the MCRG approach, make some remarks on the computational aspects and give some results.

3.1 MCRG Formalism

The following equations are intended only to expose the philosophy and notation; for a fuller discussion, see the review by Swendsen[6]. Imagine we have an Ising model with spins $\sigma = \pm 1$ at each site of a lattice, and with Hamiltonian

$$\mathcal{H} = \sum_{\alpha} K_{\alpha} S_{\alpha} \quad . \tag{3.1}$$

Here S_{α} are a set of interaction terms with e.g. S_1 the sum of nearest neighbour interactions over the lattice; K_{α} are the corresponding coupling constants. Thermodynamic quantities and correlation functions are calculated by averaging over all configurations of the spins $\{\sigma\}$, with probability distribution $\exp(\mathcal{H})$. For $K_1>0$ and in the absence of competing interactions we expect the model to undergo a continuous phase-transition at which the correlation length ξ would diverge in an ideal infinite system. The resulting singularities in thermodynamic quantitites at this critical point are smeared by finite size effects in numerical simulations.

The basic idea in the renormalisation group is to block local clusters of the original spin variables $\{\sigma\}$ into block spin variables

$\{\sigma'\}$. The renormalisation group transformation gives the effective couplings K' for the block spin interactions S'_α, in terms of the original couplings K_α. After n+1 blockings, we obtain $K_\alpha^{(n+1)}$ as functions of $K_\alpha^{(n)}$. The corresponding flow is indicated schematically in figure 3. This indicates the critical surface (the set of points where

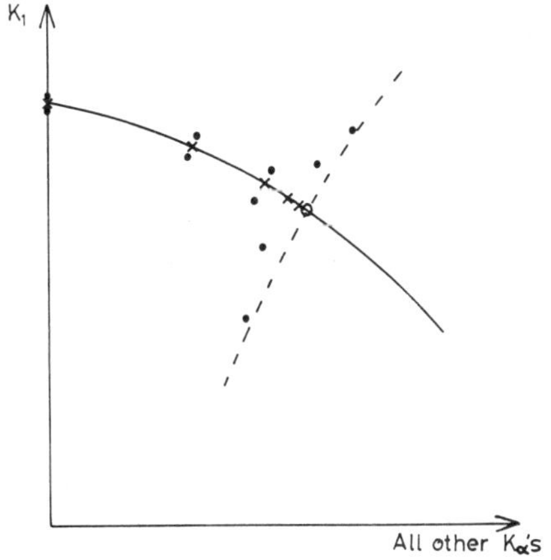

Figure 3. Schematic diagram of RG flows in the many-dimensional space of coupling constants, indicating the transient in to the fixed point starting from the critical value K_1^c, and the instability of the fixed point to perturbations out of the critical surface.

the phase transition occurs, with $\xi = \infty$) and the expected fixed point of the transformation, whose stability properties control the flow within and away from the critical surface. At the fixed point the theory is scale invariant on all length scales in the sense that any correlation function of the original spins on the original lattice equals the corresponding correlation function of the block spins on the block lattice. For any other point on the critical surface, such as the nearest neighbour Ising model at the critical coupling K_1^c, there is a transient effect as the coupling approaches the fixed point so that scale invariance emerges only at large distances. The corresponding transient behaviour gives corrections to scaling also for a system in which ξ is large but finite. Field theory calculations suggest that there is a particularly slow transient (and hence important correction to scaling) associated with the ϕ^4 operator. An important aim of MCRG is to expose this transient in the coupling constant flow.

The MCRG procedure we follow involves simulating the Ising model with nearest neighbour coupling constants on a range of lattice sizes differing by a power of 2. From an equilibrium ensemble of original spin configurations one generates an ensemble of block spin configurations; in our case the block spin for a 2^3 cell is given by the majority rule with random tie-breaker. The set of configurations of block spins is then an equilibrium ensemble for the block Hamiltonian $\mathcal{H}'$, whose couplings are neither known nor required, since the average over the ensemble estimates any required average of the block spins.

The critical coupling K_1^c is estimated by comparing expectation values of the block spin operators on "large" (L) and "small" (S) lattices at blocking levels such that the effective lattice size is the same for each:

$$\langle S_\alpha^{(n)} \rangle_L = \langle S_\alpha^{(n-m)} \rangle_S \quad . \tag{3.2}$$

In the comparison, finite size effects are cancelled and the best estimate in our case arises from comparing 64^3 data ("L") for n=5 and 32^3 data ("S") for n-m = 4. The instability of the fixed point to deviations of K_1 from K_1^c gives in practice a factor of 6 in sensitivity for a factor of 2 change in scale size. If K_1 is near K_1^c, we estimate the deviation $K_1 - K_1^c$ by

$$\langle S_\alpha^{(n)} \rangle_L - \langle S_\alpha^{(n-m)} \rangle_S = \left(\frac{\langle S_\alpha^{(n)} \rangle_L}{\partial K_1} - \frac{\langle S_\alpha^{(n-m)} \rangle_S}{\partial K_1} \right) (K_1 - K_1^c), \tag{3.3a}$$

using

$$\frac{\partial \langle S_\alpha^{(n)} \rangle_L}{\partial K_1} = \langle S_\alpha^{(n)} \, S_1^{(0)} \rangle_L - \langle S^{(n)} \rangle_L \langle S_1^{(0)} \rangle_L . \tag{3.3b}$$

The critical exponents y_a are given in terms of the eigenvalues λ^a of the stability matrix $\partial K_\alpha^{(n+1)}/\partial K_\beta^{(n)}$ evaluated at the fixed point K^*:

$$\lambda^a = 2^{y_a} . \tag{3.4}$$

We use Swendsen's approach for estimating the λ^a's, by solving equations of the form

$$\frac{\partial \langle S_\alpha^{(n)}\rangle}{\partial K_\beta^{(n-1)}} = \sum_\gamma \frac{\partial K_\gamma^{(n)}}{\partial K_\beta^{(n-1)}} \frac{\partial \langle S_\alpha^{(n)}\rangle}{\partial K_\gamma^{(n)}} \tag{3.5}$$

for the stability matrix, using correlation functions as in (3.3b). Modulo the short range finite size corrections in the RG transformation itself, best estimates are obtained by taking the highest n values on the largest lattice, so that the effective Hamiltonian is as close as possible to the fixed point; the slow transient towards the fixed point again plays an important role here.

3.2 Computational Aspects

With an EXOR random number generator, we obtain six million update attempts per second (6MHz). We actually used a NAG routine which ran at 2.7MHz. The RG blocking is performed every fourth sweep through the lattice, bringing the effective rate down to 1.2 MHz; this is certainly a too frequent blocking, given the long time correlations at large distances. Statistics are improved by shifting systematically the choice of origin for each blocking.

The results reported are for 32, 16, 4 and 2 million sweeps on 8^3, 16^3, 32^3 and 64^3 lattices respectively at each K_1 value. The entire calculations used some 500 hours of DAP time.

3.3 Results

The first runs were done at $K_1 = 0.22169$ (the old series expansion estimate for K_1^c) and then at $K_1 = 0.22161$. Using (3.3) we then estimated $K_1^c \simeq 0.22166$, where a third run was then done. The resulting deviations $K_1 - K_1^c$, obtained by comparing expectation values of the nearest neighbour operator (i.e. for $\alpha = 1$ in (3.3)) for various lattice sizes and blockings, are given in table 1. The final column gives the estimates for K_1^c by taking the statistical means from the three different runs. The closeness of the final results from the larger lattices

Table 1: Estimates for $(K_1^c - K_1)\times 10^6$ obtained from runs at three different K_1 values, comprising large (L) and small (S) lattices at successive RG blockings. The final column contains the corresponding best estimates for K_1^c obtained by averaging over the results from the three runs.

		K_1 =	0.22161	0.22166	0.22169	
L	S	RG	change(err)	change(err)	change(err)	mean(err)
64	32	1	-528(11)	-423(10)	-365(9)	0.222086(5)
64	32	2	2(8)	57(7)	81(7)	0.221606(4)
64	32	3	-32(9)	28(8)	54(8)	0.221636(4)
64	32	4	-50(12)	16(10)	42(11)	0.221650(6)
64	32	5	-58(13)	17(10)	38(15)	0.221652(7)
64	16	1	-144(5)	-79(6)	-43(6)	0.221744(3)
64	16	2	-12(4)	43(7)	70(7)	0.221621(3)
64	16	3	-38(6)	21(8)	49(9)	0.221644(4)
64	16	4	-52(6)	14(9)	39(12)	0.221656(4)
64	8	1	-53(4)	6(6)	35(6)	0.221659(2)
64	8	2	-20(5)	40(7)	66(8)	0.221626(3)
64	8	3	-40(6)	28(8)	52(11)	0.221643(4)
32	16	1	-556(11)	-492(6)	-447(6)	0.222147(3)
32	16	2	45(12)	87(6)	120(7)	0.221571(4)
32	16	3	-5(15)	36(7)	69(9)	0.221622(5)
32	16	4	-34(21)	5(10)	44(14)	0.221651(7)
32	8	1	-97(8)	-42(4)	-11(3)	0.221702(2)
32	8	2	43(10)	92(4)	122(4)	0.221568(2)
32	8	3	1(14)	51(6)	84(6)	0.221607(4)
16	8	1	-519(7)	-435(5)	-420(11)	0.222107(3)
16	8	2	185(8)	263(7)	282(16)	0.221409(5)
16	8	3	103(11)	191(9)	204(21)	0.221484(6)

indicates that any extrapolation allowing for the slow transient neglected in (3.3) is likely to be small compared with the statistical error in the results. Our current best estimate for the critical coupling is

$$K_1^c = 0.221656(5) \tag{3.6}$$

The errors in table 1, on which this result is based, are the standard deviations in the mean from the data divided into eight successive blocks; even with the change of origin in the blocking procedure, the long time correlations on the 64^3 lattice imply that the successive data points are barely independent and the standard deviation in (3.6) is certainly not overestimated. The result should be compared with recent series expansion estimates (allowing for the effect of the leading correction to scaling): 0.221655(10)(ref.8), 0.221662 (ref.9), 0.221655(5)(ref.10); it appears to be also in good agreement with results from the Ising model machine at Santa Barbara [11].

The estimates for the leading critical exponent $y_1 \equiv 1/\nu$ from (3.5) in the space of operators even in the spins is shown in table 2, for the various lattices. The data shown are from the run at K_1=0.22166. A set of seven successive values in a column comes from diagonalising the stability matrix with an increasing number of operators (four two-spin and three four-spin, maximum). The successive sets in a column are for increasing blocking levels n in (3.5). The transient effect associated with the approach to the fixed point is evident in the table. There is also an apparent finite size effect in the renormalisation group transformation itself (compare horizontally across the table, at the highest blocking values); this may be in fact a signal that more than seven operators are necessary. If we take as a bench mark the starred value obtained by comparing 8^3 and 4^3 lattices in the 64^3 data, it gives ν = 0.628 with a statistical error of less than 1 in the third figure. Correcting for (i) the interpolation (from the other two data runs) to the best estimate for K_1^c, (ii) the apparent finite size effect and (iii) the extrapolation for the slow transient, our current estimate is

$$\nu = 0.629(5) \quad . \tag{3.7}$$

It is clear that in this case the error is determined mainly by the last two systematic effects. The result is again in good agreement with the recent estimates 0.628 - 0.633 for the simple cubic lattice [10], 0.631(3) for the body centred cubic lattice [12], 0.630(2) from field

Table 2: Estimates for the exponent $1/\nu$ obtained at K = 0.22166 using eq. (3.5).

64^3	32^3	16^3	8^3
1.36631(70)	1.36483(54)	1.36552(30)	1.37579(25)
1.42641(72)	1.42877(71)	1.43714(22)	1.45977(25)
1.42705(59)	1.42929(78)	1.43760(30)	1.46095(27)
1.42602(63)	1.42745(75)	1.43397(30)	1.45279(46)
1.42590(61)	1.42725(75)	1.43330(31)	1.44804(41)
1.42540(65)	1.42633(71)	1.43125(35)	1.43928(47)
1.42519(72)	1.42581(82)	1.43027(35)	1.43715(44)
1.49366(86)	1.49410(70)	1.51365(36)	1.59218(29)
1.52440(68)	1.53266(80)	1.56561(59)	1.68669(25)
1.52459(82)	1.53259(90)	1.56624(59)	1.69557(26)
1.52103(93)	1.52713(94)	1.55728(56)	
1.52051(92)	1.52566(91)	1.54815(60)	
1.51929 (92)	1.52248(91)	1.53576(78)	
1.51888(93)	1.52182(91)	1.53384(76)	
1.54485(126)	1.56429(120)	1.65851(68)	
1.57677(98)	1.60993(111)	1.74601(54)	
1.57649(97)	1.61030(119)	1.75447(52)	
1.57046(103)	1.60063(138)		
1.56881(96)	1.59064(144)		
1.56557(88)	1.57799(126)		
1.56514(84)	1.57599(135)		
1.58513(314)	1.68662(296)		
1.62846(290)	1.77281(298)		
1.62850(297)	1.78104(310)		
1.61896(259)			
1.60845(236)			
1.59435(209)			
*1.59220(204)			
1.69654(922)			
1.78028(867)			
1.78814(871)			

theory calculations [13] and 0.635(5) from finite size scaling calculations [14].

We have similarly calculated the leading exponent for the odd operators; our current estimate for the conventional exponent η is

$$\eta = 0.028(5)$$

This is in similar agreement with other estimates.

The picture for the second largest eigenvalue in the even sector, the correction to scaling exponent, is less clear. We obtain $\omega \simeq 1.1$ with error bars of roughly 15%, compared with other estimates of 0.8 - 0.85. In the odd operators, the second exponent is actually positive; one would expect it to be associated with the redundant operator [15)] ϕ^3 in field theory language. This hypothesis is testable since its value ($\sim$ 0.4 in our case) would depend on the RG prescription. The redundant operator corresponding to $\phi\delta\mathcal{H}/\delta\phi$ may also be confusing the determination of the correction to scaling exponent in the even operators.

In conclusion we note that the above determination of K_1^c is statistics limited; the accuracy of exponents, which do not in the above approach involve comparison of runs on different lattice sizes, is largely limited by systematics. Details of the above will appear in ref.(5).

4. Hadron Mass Calculations

We have set up the standard core of programmes for quenched lattice QCD [16)] on the DAP. Pure gauge SU(3) programs, for an 8^4 lattice and with the standard Wilson action, have been written using both the Metropolis algorithm and the modified heat bath method [17)]. Both programs run at approximately 200μ sec/link update. The quark propagator for a given background gauge field is calculated by a Gauss-Seidel routine, for Wilson fermions on an $8^3\times 16$ configuration formed from two copies of an 8^4 configuration. One of the major problems of implementing SU(3) on the DAP is squeezing all the variables for one lattice site - including of course all working space - on to the 4K RAM which each PE has in our machine. This is achieved by using two-byte integer storage for the link variables and three-byte floating point arithmetic for SU(3) multiplies and the quark Green functions; these particular word lengths appear adequate given the statistical and systematic effects in our calculations to date. The calculation of the rest frame hadron propagators at each of the sixteen time slices is also done on the DAP.

We have so far done extensive runs at bare couplings $1/g^2=1.0$ and $1/g^2 = 0.95$. Ref. 18 reports the results of an analysis of the pion propagator at $g^2 = 1.0$ using 40 Metropolis configurations obtained on the DAP at Queen Mary College. The conclusions of that paper are:

(i) The exponential decay of the π propagator does not emerge at this g^2 value until at least 7 or 8 lattice spacings from the origin.

(ii) Finite size effects are still significant because there is still a correlation between the pion mass from a given configuration and the average of the Wilson loops of length 8 round the lattice in the spatial directions for that configuration. The spread in mass value is however much less than in ref. 19.

(iii) The comparison with other estimates of the pion mass is summarised in figure 4; the significantly larger values obtained in earlier work are largely understandable in terms of finite size effects.

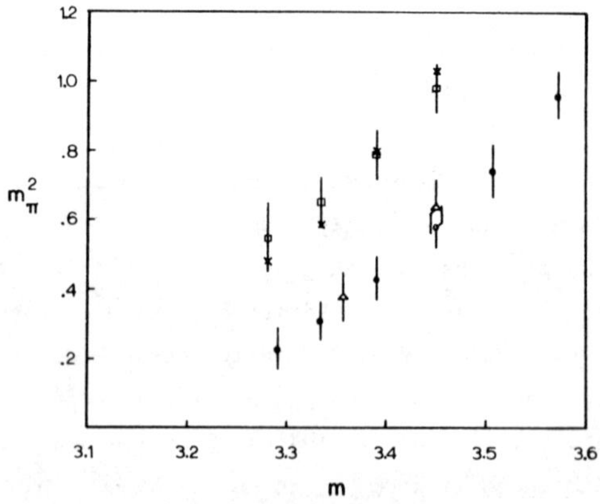

Figure 4. Pion mass-squared as a function of quark mass parameter at $g^2 = 1.0$:● this work (40 configurations on $8^3\times16$): x Hamber and Parisi (ref.16): □ Fucito et al (ref.16): Δ Bernard et al (ref.22).

(iv) From our results in figure 2 we estimate the critical value of the hopping parameter, where the π mass vanishes, to be $K_c = 0.157(1)$.

These calculations have been repeated for an ensemble of 40 heat bath configurations [20]. The statistical quality of these configurations is much better and the general picture is as above. The results for the pion mass are a little lower than our results in figure 4, but within a standard deviation. K_c is estimated to be 0.156(1).

In recent work [21] we have calculated the particle propagators which can be formed from all local covariant $q\bar{q}$ and qqq operators (except for the second Δ operator $(\psi C \sigma_{\mu\nu} \psi)\psi)$. We have analysed 16 configurations at $g^2 = 1/0.95$ and 24 at $g^2 = 1.0$. The results for two pion $(\bar{\psi}\gamma_5\psi$, $\bar{\psi}\gamma_4\gamma_5\psi)$, two rho $(\bar{\psi}\gamma_i\psi, \bar{\psi}\gamma_4\gamma_i\psi)$, two proton $((\psi C\gamma_5\psi)\psi)$ and $(\psi C\gamma_4\gamma_5\psi)\psi)$ and delta $((\psi C\gamma_\mu\psi)\psi)$ operators are shown in fig. 5.

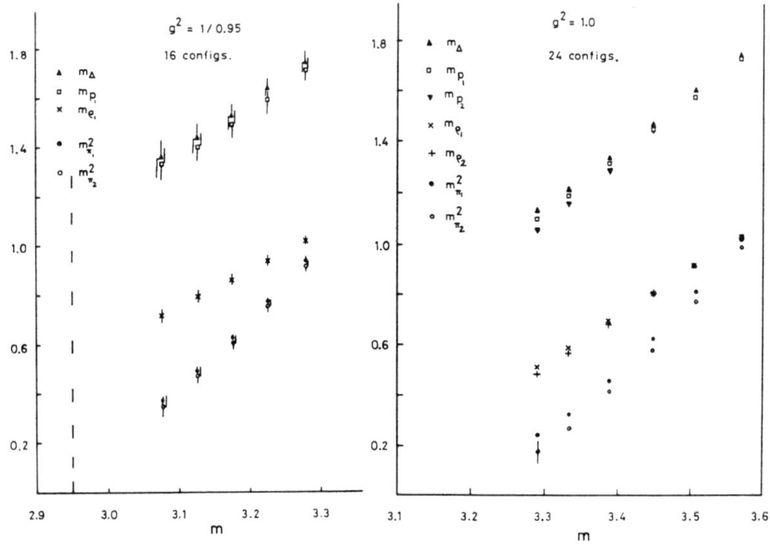

Figure 5. Hadron masses at (a) $g^2 = 1/0.95$ and (b) $g^2 = 1.0$. The error bars are based on the standard deviation measured from two blocks of 8 configurations. In (b) all error bars are comparable to that shown for π2.

The following remarks reflect our present conclusions from the data.

(i) We used in these calculations fixed (Dirichlet) boundary conditions on the quark propagator in time direction (following ref.22). Although edge effects are not negligible, the emergence of the exponential asymptotic decay of a propagator did appear to be better exposed than with the periodic boundary conditions of our previous work.

(ii) The quality of data can be gauged from the plots of effective mass as a function of time slice in figure 6. The results for mesons are very encouraging but even at the larger g^2 value the baryons are

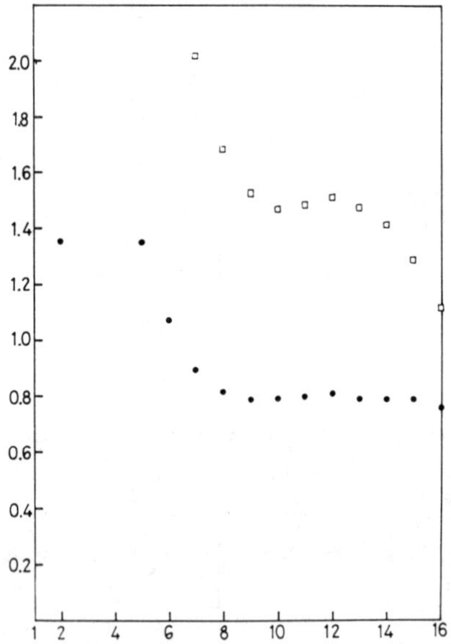

Figure 6. Effective mass plots, $m(n_4) = -\ell n[\Delta(n_4)/\Delta(n_4-1)]$ (following ref.22) as a function of time slice n_4, for pion (•) and proton (□). The data are from 16 configurations at $g^2 = 1/0.95$, for hopping parameter $K = 0.1575$.

not yet very convincingly exposed. The apparent oscillation in the baryon masses could be a consequence of the boundary source or fluctuations. (Note that the origin was placed at time slice 3 after various tests as a compromise between the competing demands of mass and residue estimation.)

(iii) At $g^2 = 1/0.95$, the masses obtained from the two different channels for π, ρ and proton are compatible within statistical errors (taken as the standard deviation on a block of 8 configurations). The extrapolation of m_π^2 and m_ρ as linear functions of the quark mass parameter $m(\equiv 1/2K)$ appears satisfactory. We estimate $K_c = 0.1695(7)$ and the ρ mass in units of the lattice spacing a to be

$$m_\rho a = 0.52(4) \tag{4.1}$$

Using the physical value of the ρ mass (770 MeV) yields a lattice spacing

$$a = 0.133(10)\text{fm}; \qquad a^{-1} = 1.48(12)\text{GeV} \quad . \tag{4.2}$$

This estimate of lattice spacing in fermi is significantly smaller than previous estimates. In the case of other estimates from hadron masses, the discrepancy would appear to arise from our generally lower mass values. However, bearing in mind that the finite size effect at $g^2 = 1.0$ discussed in (vi) below appears to reduce the estimate for the ρ mass below the true value, (4.2) may be an underestimate of the lattice spacing in fermi. As regards string tension measurements of a, present data [23] suggests that they could decrease as larger lattices are studied.

(iv) The baryon masses in figure 5(a) present two problems. The p/ρ mass ratio is far too large at roughly 2 (and this is in line with the estimates of refs. (24) and (22)). Further, although there is a Δ-p mass splitting, at less than 100 MeV it is far too small. These are typical features of the strong coupling limit. Furthermore, a box of size just over 1 fermi may be adequate for mesons but is probably inadequate for baryons.

(v) As g^2 is decreased to 1.0, we expect the hadron to occupy a correspondingly greater number of lattice sites. The main feature of the data summarised in figure 5(b) is the existence of significant finite size effects. First, although error bars for the π mass from the two operators overlap, the masses are correlated and the corresponding mass difference is statistically non-zero by several standard deviations. That the discrepancy in the data must be a finite size effect is underlined by the following remark. The $\pi_1 = \bar{\psi}\gamma_5\psi$ operator yields the largest meson propagator because it is the sum of squares of components of the quark propagator. Therefore in an infinite system with exponential decay with distance, no other propagator can yield a mass lighter than π_1 does; this is manifestly violated by $\pi_2(=\bar{\psi}\gamma_4\gamma_5\psi)$ data. Linear extrapolations suggest

$$K_c^{(\pi_1)} = 0.1560(12) \ ; \qquad K_c^{(\pi_2)} = 0.1546(10) \qquad (4.3)$$

In one respect π_2 may have smaller finite size effects, since its residue at small pion mass behaves as $m_\pi^2 f_\pi^2$ (c.f. f_π^2 for π_1).

(vi) A further problem is that the ρ masses are almost degenerate with π_1. (This is masked in figure 5(b) because we plot m_π^2 and m_ρ. It may again be significant that the lighter of the two ρ particles is the one with smaller residue). One therefore cannot with these data make a convincing estimate of the lattice spacing a at $g^2 = 1.0$. However if we presume K_c is in the range 0.154 - 0.157, a linear extrapolation of m_ρ yields a lattice spacing in the range

$$a = 0.094(18) \text{ fm} \ ; \qquad a^{-1} = 2.1(4) \text{ GeV} \qquad (4.4)$$

This value for a is again much smaller than previous estimates but it and (4.2) are crudely in line with continuum scaling arguments from asymptotic freedom.

(vii) The transverse lattice dimension of roughly 0.75 fm is presumably too small, both for mesons and baryons, (even with the range of quark masses used here) and it is hardly surprising that the problems with baryons noted at $g^2 = 1/0.95$ persist at $g^2 = 1.0$. A Δ-p splitting does appear to be opening up but this effect is in parallel with a mass splitting between two proton operators which is presumably a finite size effect.

In summary, we believe that the above provides strong evidence for the conclusion [25] that larger lattices are required, preferably with improved actions, in order to obtain a clear window of continuum physics.

Further details of the hadron mass calculations will be published in ref.(21). We stress again that mass estimates quoted above and the MCRG results of section 3 should be regarded as preliminary.

Acknowledgements

I acknowledge the essential contribution made by collaborators in the work reported here; E. Marinari, B.J. Pendleton, F. Rapuano, R.H. Swendsen, K.G. Wilson and particularly K.C. Bowler and G.S.Pawley.

References

1. For further information on the DAP and its software see e.g. R.W. Hockney and C.R. Jesshope, Parallel Computers (Adam Hilger Ltd. Bristol, 1981); G.S. Pawley and G.W. Thomas, J.Comp.Phys. 47 (1982) 165.
2. G.S. Pawley and G.W. Thomas, Phys. Rev. Lett. 48 (1982) 410; M.T. Dove and G.S. Pawley, J. Phys. C. in press; G.S. Pawley and M.T. Dove, Chem. Phys. Lett. 99 (1983) 45.
3. K.C. Bowler in Proceedings of the Three-day In-depth Review on the Impact of Specialized Processors in Elementary Particle Physics, Padua, March 1983.(University of Padua, 1983).
4. I.O.Stamatescu, Max Planck preprint MPI-PAE/PTh 82/81.
5. G.S. Pawley, R.H. Swendsen, D.J. Wallace and K.G. Wilson to be published.
6. R.H. Swendsen, in Real Space Renormalisation, eds. T.W.Burkhardt and J.M.J. van Leeuwen (Springer-Verlag, 1982)57.

7. Reviews of series expansion results and the importance of corrections to scaling can be found in Phase Transitions (Proceedings of Cargèse Summer School 1980), eds. M. Levy, J.C. Le Guillou and J. Zinn-Justin (Plenum, 1982).
8. J. Zinn-Justin, J. de Physique 42 (1981) 783 and private communication.
9. D.S. Gaunt, in ref. 7 and private communication.
10. J. Adler, Technion preprint 82-49.
11. R.B. Pearson in Proceedings of Les Houches Workshop, March 1983, Phys. Repts., to be published.
12. B.G. Nickel, in ref. 7.
13. J.C. Le Guillou and J. Zinn-Justin, Phys.Rev. Lett. 36 (1976) 1351 and Phys. Rev. B21 (1980) 3976.
14. C.J. Hamer, J. Phys. A16 (1983) 1257.
15. F.J. Wegner in Phase Transitions and Critical Phenomena, Vol.VI eds. C. Domb and M.S. Green (Academic Press, 1976) 34.
16. Early references are H. Hamber and G. Parisi, Phys. Rev. Lett. 47 (1981) 1792; D. Weingarten, Phys. Lett. 109B (1982) 57; E. Marinari, G. Parisi and C. Rebbi, Phys. Rev. Lett. 47 (1981) 1795. See also F. Fucito, G. Martinelli, C. Omero, G. Parisi, R. Petronzio and F. Rapuano, Nucl. Phys. B210[FS6] (1982) 467.
17. N. Cabibbo and E. Marinari, Phys. Lett. 119B (1982) 387.
18. K.C. Bowler, G.S. Pawley, D.J. Wallace, E. Marinari and F. Rapuano, Nucl. Phys. B220 FS8 (1983) 137.
19. G. Martinelli, G. Parisi, R. Petronzio and F. Rapuano, Phys. Lett. 122B (1983) 283.
20. K.C. Bowler and B.J. Pendleton, Nucl. Phys. B, to appear.
21. K.C. Bowler, G.S. Pawley and D.J. Wallace, to be published.
22. C. Bernard, T.Draper and K. Olynyk, Phys. Rev. D27 (1983) 227.
23. See e.g. E. Pietarinen, Nucl. Phys. B190 FS3 (1981) 349; R.W.B. Ardill, M. Creutz and K.J.M. Moriarty, Phys. Rev. D27 (1983) 1956.
24. D. Weingarten, ref. 16 and Indiana preprint IUHET-82 (1982).
25. P. Hasenfratz and I. Montvay, Phys. Rev. Lett. 50 (1983) 309.